STONE BUILDINGS

Praise for Patrick McAfee's
previous book, *Irish Stone Walls*

'It's splendid. The wealth and variety of the Irish stone
tradition is displayed as never before. A source of joy
... written with feeling, even with passion.'

MAURICE CRAIG

'Pat McAfee communicates with a clarity that is based
on knowledge. This book should be on the desk of
every local authority architect, engineer and planner.'

**PROF. LOUGHLIN KEALY,
HEAD OF THE SCHOOL OF ARCHITECTURE, UCD**

Patrick McAfee is an expert stonemason. Born and brought up in Dublin where he served an apprenticeship to his father, he worked for a number of years in Australia and studied traditional methods of working with stone and lime mortars at the European School of Conservation at San Servolo in Venice. He now divides his time between workshops on stone and lime, and consultancy work around the country.

Also by Pat McAfee

IRISH STONE WALLS, History, Building, Conservation

STONE
BUILDINGS

Conservation ▪ Repair ▪ Building

Patrick McAfee

THE O'BRIEN PRESS
DUBLIN

First published 1998 by The O'Brien Press Ltd
12 Terenure Road East, Rathgar, Dublin 6, Ireland.
Tel: +353 1 4923333; Fax: +353 1 4922777
E-mail: books@obrien.ie
Website: www.obrien.ie
Reprinted 2001, 2004, 2010.

ISBN 978-1-84717-210-5

A catalogue record for this title is available from The British Library

4   5   6   7   8   9   10
10   11   12   13

This publication is sponsored by
The Heritage Council / *An Chomhairle Oidhreachta*

Editing, layout, typesetting, design: The O'Brien Press Ltd
Printed by Lightning Source UK Ltd.

Cover photographs
Back Flap (top) Ballitore, County Kildare'
(bottom) Limewashing St Peter's Church, Cork City.
Back cover (top) the author pointing ashlar work;
(bottom) Drimnagh Castle, Dublin.
Front cover (top) Newtown Castle, Ballyvaughan, County Clare;
(bottom, from left) Vernacular farm building, Dublin;
Traditional stone house in Collon, County Louth; Custom House, Dublin.
Front flap (from top) Trinity Presbyterian Church, Cork;
Irish stonecutter working with compressed air in Florence, Italy;
Stone-walling workshop, Grange Castle, County Kildare.

Stonemasons at work, An Spidéal, County Galway, c.1939.

DEDICATION

For my parents, Hugh and Greta

Acknowledgements

To my family Ann, Brian and Barbara, my parents Hugh and
Greta, and to my brother and family Gerard, Margaret and Sharon.
To the following members of FÁS: Eamon Rapple, Jane Forman, Pat
Kelly Rogers, Alan Farrell and James Farrell.
I would also especially like to thank Amanda Wilton, Sile O'Kelly,
James O'Callaghan, James Howley and the directors and staff of The
O'Brien Press.

Kilconnell Friary, County Galway.

CONTENTS

FOREWORD

In this, his second book, Pat McAfee again shows that he is an Irish craftsman in the finest sense of that word. He is not only a skilled stonemason, but in the style of a true master of his craft he understands Irish traditional buildings in their historical, scientific and global contexts, and describes them with confidence and pride.

As stonemason, he describes the practical essentials of his trade – stonecutting, lime mortars and renders, the tools, the methods. He has an understanding of the scientific composition and qualities of the materials he works with. Much more than this, however, with pride in his inheritance, he places the humblest house in its global context, with descriptions of classical details, international charters and principles of conservation.

His expertise and understanding, combined with the ability to communicate to others, both in his teaching and through his books, have brought the traditions and building skills of the past back from near oblivion. We are taught to look around at the simplest structures we take for granted, and are in danger of losing. We begin to see them with a real understanding of their importance, as buildings, which, if treated with respect and care, can be a viable part of our future, and not just occasional reminders of the past.

The Council appreciates the contribution Pat continues to make to our heritage, and is happy to sponsor this book and be associated with him and his work.

Freda Rountree
Chairperson
The Heritage Council

An
Chomhairle
Oidhreachta

The
Heritage
Council

INTRODUCTION

This is a simple book mostly about old buildings built of stone and lime mortar. Like *Irish Stone Walls* it attempts to take the reader through traditional ways of working with materials, this time in relation to stone buildings. Why is this important? Because by studying past practice we can begin to understand and respect how old buildings were constructed and how we should conserve and repair them in the present. Without our understanding and respect they will continue to disappear or they will be repaired and modernised in ways that make them indistinguishable from modern buildings. This is happening for many reasons, but mainly because of the general belief that it is either impossible or very expensive to make old buildings habitable to a reasonable standard.

People have a fascination with stone buildings and much has been written on their architectural history, describing the lives of owners, the history of their times, and detailing their architectural styles and influences. But very little is to be found about the materials and skills used in their *construction* and on how we should now conserve and repair them. As a result, myths abound that separate us even further from fact. Both professionals and house owners are concerned with these buildings, and this book attempts to provide for the practical needs of both groups. Readers should dip in and select those elements that are of use and relevance to their own needs. Skills are outlined in detail, where possible, to help people understand what they are looking at in an old building, and to repair what needs to be repaired.

Influences came and went but basic ways of working were handed down for generations and even centuries, and these skills were adapted to changing times and styles. These ways of working – unrecorded and guarded often as mysteries – are only to be found in the buildings themselves. To read what is there, to understand how it got there and what we should now do to repair and conserve it is the purpose of this book. Photographs and freehand drawings have been included as visual guides to help you identify what you see and to complete any necessary intervention to buildings with skill and sensitivity. To describe a practice is always interesting and sometimes difficult compared to explaining by doing. I write from a background of practical experience and I hope I have managed to explain the details clearly here. It seems to me that there is no end to what we can learn about these buildings. May you enjoy this book, and come to look at and after old buildings not as mysteries but as places to live and enjoy, and which, because they are old, will at times need a little extra care, know-how and sympathy.

Pat McAfee
June 1998

THE CLASSICAL INFLUENCE

There has been no greater influence on western architecture than the legacy of Greece and Rome. This classical influence has reached us through the Renaissance and the subsequent stages of mannerism, baroque, rococo, neo-classicism, including Greek revival and so on. It has spread far and wide, and persisted to the present century. In its strictest sense it is based on the classical orders with their easily recognisable elements, particularly their capitals.

A more lenient interpretation of 'classical' is often applied to simpler buildings lacking much or nearly all of the detail of the orders but still with a classical sense of harmony, symmetry and proportion. This classical influence is seen in the grandest of buildings, but also very often in the simplest buildings too. A grasp of classical proportions is essential if we are to tackle the conservation of buildings, large and small. Sometimes such influence is not immediately obvious: in many of the Georgian houses of Dublin all the splendour is seen internally, executed in plasterwork, furniture and fittings with only a classical door surround used externally in an otherwise plain brick façade – but the proportions are classical. How many people are aware that the skirting at the base of an interior wall, and the dado (chair) rail, architrave, frieze and cornice represent elements of the pedestal and entablature found in classical architecture? Externally, many buildings have plinths, quoins, parapets – again, all derived from classical architecture.

The knowledge of the classical orders was once taught as an essential part of the training process of the architect, stonecutter, plasterer, carpenter, joiner, cabinet-maker, painter, blacksmith. The classical influence was so complete that one is left with the feeling that to design or construct anything outside this would have taken a great conscious effort. It influenced both the buildings and their decoration and furnishing.

The conservation and repair of even humble classical buildings requires at least a basic understanding of the classical orders.

The orders began with classical Greek architecture (7th century BC – 1st century AD) based on the column and beam. The Greeks developed the first three orders.

1 Doric

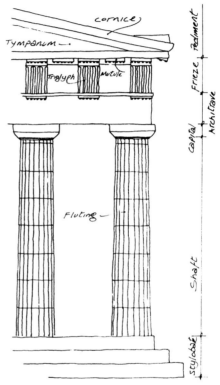

Above: Plan of a Doric column showing fluting and dowel.

Right: Doric temple – the Parthenon in Athens, which is considered to be the most beautiful building in the world. Note the elements and proportions in this elevation.

2 Ionic

Ionic columns were more intricately decorated than the plain Doric.

3 Corinthian

Corinthian columns are richly
decorated with acanthus
leaves.

The Romans (5th century BC – 4th century AD) also used these first
three orders but added two more.

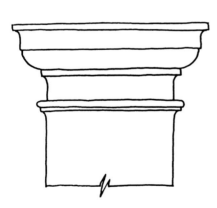

5 Composite

Composite (Renaissance example). The composite capital
combines elements of both Ionic
and Corinthian design.

4 Tuscan

Tuscan (Renaissance example
shown). The shaft is not fluted.

The Romans inherited the Tuscan order from the Etruscans (as well as the arch) and they developed the Composite by amalgamating both the Ionic and the Corinthian orders.

There are therefore five orders. The Renaissance applied a systematic approach to their study and use. This is well expressed in the drawing, opposite, from *Normand's Orders of Architecture* by R.A. Cordingley, showing their Renaissance interpretation. Although fault may be found in expressing them in such a rigid way, at the same time much useful information about proportions is to be seen here which can be easily memorised and adapted.

It can be seen from this drawing that the base diameter of the column shaft of each of the orders is the module used to determine the proportions of all the other elements. This module is also used elsewhere in the arrangement, or set out, of classical buildings. We can see that the Tuscan and Doric orders are squat, while the Corinthian and Composite orders are long and elegant, with the Ionic order in between.

On three-storey formal classical buildings it is usual to place the Doric or Tuscan order on the ground floor, the Ionic order on the first floor and the Corinthian or composite order on the second floor.

As stated earlier, the orders are based on the column and beam. A **column** has three elements

·   Base
·   Shaft
·   Capital.

The original Greek Doric column had no base. The capital of each order is instantly recognisable. The shaft of all the orders is tapered from base to top but not in a straight line. The Greeks recognised that a column with straight lines would look waisted (concave/hollow), they therefore cut the shafts of the column to have a slight swelling, this is called **Entasis**.

Column surfaces (except the Tuscan) are sometimes cut with concave vertical channels. This is called **Fluting**.

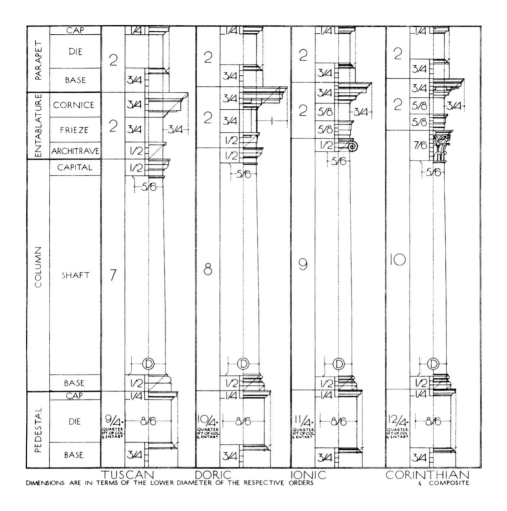

DIMENSIONS ARE IN TERMS OF THE LOWER DIAMETER OF THE RESPECTIVE ORDERS

A method of drawing the Renaissance orders by R.A. Cordingley in *Normand's Parallel of the Orders of Architecture*, Tiranti 1959 (6th edition). Much useful information is shown here for design of buildings, shop fronts and even entrance walls and piers.

On top of the capital of the column is the **Entablature** which is composed of three elements:

·   Architrave

·   Frieze

·   Cornice.

The architrave is the beam which carries the overhead weight. The frieze is sometimes decorated and the cornice projects, giving protection to the building.

The triangular gable end of the building is called the **Pediment**. Pediments occur elsewhere, for instance on a smaller scale over windows and doors.

The orders originated from temple architecture and their columns sat on a platform composed of three steps called a **Stylobate**.

The Romans used the orders for buildings other than temples and introduced the pedestal under the column and the parapet over the entablature. This became very popular from the Renaissance onwards. Both the **pedestal** and the **parapet** are composed of three elements,

·   Base or plinth

·   Die or dado

·   Cap or cornice.

A shallow pier attached to and projecting from a wall, usually rectangular in cross-section, and imitating a column is called a **Pilaster**.

When projections are more than half a column in width they are called **Engaged columns**.

Columns do not always start at ground-floor level but sometimes start at first-floor or other levels. When columns with entablature/pediment extend up the face of a building for more than one storey they are referred to as a **Giant order**.

Other elements are strongly associated with classical architecture such as the

- Semi-circular arch
- Vault
- Dome.

The domed Pantheon in Rome is one of the finest Roman buildings remaining today.

Concrete made from lime and aggregate (sand, stone and brick) plus pozzolana (see Chapter 12, Lime) was developed and fully exploited by the Romans in many of their constructions including domes.

The following **mouldings** are to be seen on column bases, cornices and so on:

- Torus
- Fillet
- Scotia
- Cyma Recta
- Cyma Reversa
- Ovolo
- Cavetto
- Bead or Astragal
- Fascia.

Some or all of these mouldings are to be found executed in wood, plaster, metal, paint.

Mouldings and other elements are often enriched with various designs from nature or geometry. The acanthus leaf is the most common form of decoration used.

Mouldings.

Some classical buildings, particularly at ground-floor level, display large blocks of stone, quarry/rock-faced, vermiculated (worm eaten), reticulated (net-like pattern), diamond pointed, or flat smooth faced with chamfers. This style of work is referred to as **Rustication**.

Even the orders with their columns are sometimes rusticated, particularly Tuscan and Doric.

Rustication is often represented in materials other than stone, such as stucco, brick, and even paint. Projecting quoins on the corners of buildings still popular today are a form of rustication.

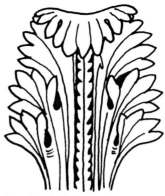

Acanthus leaf decoration.

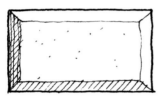

Chamfered rustication.

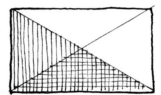

Diamond rustication.

Quarry-face rustication.

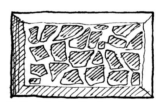

Reticulated rustication.

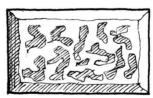

Vermiculated rustication.

Ancient and Classical Buildings

The Parthenon in Athens, Greece. Fifth-century Doric order. Considered by many to be the most beautiful building in the world.

Above: Arch of Septimius Severus in Rome.
Left: The Colosseum, Rome.

Cretan civilisation, exemplified by the palace at Knossos, had a large influence on later Greek classical architecture.

Early to Late Medieval Buildings

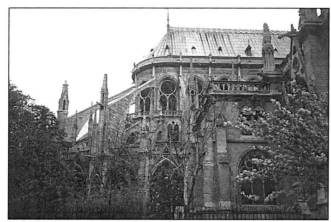

Cathedral de Notre Dame, Paris. Thirteenth-century Gothic style with flying buttresses for support.

Rock of Cashel, County Tipperary.

Above: Drimnagh Castle and moat, Dublin. Mostly fifteenth- and sixteenth-century, recently restored. The moat is still operational.

Below: Trinity Church, Glendalough. Eleventh- and twelfth-century. A nave and chancel church.

Soffit of twelfth-century romanesque arch, on a round tower, Kildare Cathedral.

Fifteenth- to Eighteenth-century Buildings

The Villa Rotonda by Palladio, sixteenth-century. A Renaissance building with world-wide influence.

Brunnelleschi's dome, Florence, fifteenth-century. Built without centering for support.

The Custom House in Dublin – eighteenth-century.

Techniques and Practices

Stonemasons working at the conservation of the Parthenon, Athens. The tools being used here have not changed in millennia.

A young Irish stonecutter cutting an arched lintel for a two-light window, Drimnagh Castle, Dublin.

Building techniques

Above: Fourteenth-century Byzantine with bond timber (Thessalonika, Greece).
Below: Thessalonika – Alatja Imaret.

CHAPTER 2

BEFORE YOU BEGIN

Many people have a dream of buying an old house that maybe needs some repair. The imagined house is located in a rural setting, beside sea, mountain, river, or lake. Some of the work, or maybe all of the work, will be carried out by themselves. It will be a labour of love, a new experience, something to be done in their spare time, at weekends, or maybe during their annual holidays.

For some, the dream becomes a reality and is the result of hard work and probably not really being aware at the beginning exactly what would be involved. For others, the reality is different: excess expenditure, too little available time, lack of energy and maybe finally putting the property up for sale without having had the pleasure of living in it.

In most property transactions there is a seller and a buyer and through ignorance either or both sides can add greatly to the final cost of the building. For instance, the seller or the buyer may unfortunately do one or more of the following:

Action: Demolish and remove outbuildings, additions and alterations because they are not needed, or are seen as taking away visually from the main building.
Negative result: The social, historical, cultural and architectural value of the site is lost.

Action: Remove the external render of the building to reveal the underlying structure of stone and advertise the property as being a traditional stone building, hoping for a higher price.
Negative result: Ingress of rain through walls seen as damp patches internally;
Increased heat loss;
Aesthetic appearance lost;
Historic material lost.

Action: Grit-blast the building after removing the external render in order to give it a clean look.
Negative result: Loss of mortar joints between stone and brick results in pointing being necessary;
Fire-skin removed off brick faces making them more porous;
Softer stones lose their faces;
Arrises (corners/edges) are rounded.

Action: Point brick and stone surfaces with a sand-and-cement mix.
Negative Result: Ability of walls to breathe is reduced resulting in damp;
Impossible to remove new pointing without damage to bricks and stones.

Action: Remove internal lime plaster to expose stone walls and create an 'old world' look.
Negative Result: Internal aesthetic appearance lost;
Historic fabric lost;
Heat loss.

Action: Remove original windows and replace with uPVC or aluminium windows to show the house has been improved to modern standards.
Negative Result: Aesthetic and historic appearance lost;
Historic fabric lost.

Action: Paint walls with impervious plastic paint to give the building a bright, fresh look.
Negative Result: Permeable walls made impermeable resulting in dampness;
Difficult to remove.

The above actions not only add considerably to the asking price of the building but also have legal consequences in the case of a structure protected under the planning laws.

There are instances of owners not only paying a contractor to have this done but then, having discovered the negative impact of their actions, paying again for it to be replaced or undone.

Excessive expenditure is incurred by the owner who wishes to over-modernise an old building and is often the reason for people

ORIGINAL HOUSE WITH BARN, WALLS, COBBLES & TREES

THE SAME HOUSE AFTER NEGATIVE INTERVENTION

FOR SALE

PMCA

referring to these buildings as being very expensive to make habitable.

Here are some common misconceptions resulting in unnecessary expense:

Action: Walls slightly leaning off plumb are plastered and rendered in order to plumb them.
Negative Result: Unnecessary excess plaster and render used to plumb walls;
Or walls demolished and rebuilt.

Action: Minor twists, undulations and imperfections on wall faces are straightened out.
Negative Result: As before;
Character of walls lost.

Action: Floors slightly off level are ripped out and new floors inserted.
Negative Result: Door heights, skirting, ceilings, etc, are all interfered with;
Danger of structural collapse.

Action: Concrete foundations are inserted under buildings that have been structurally sound for centuries and the new, intended loading is no different from what it was before.
Negative Result: Massive cost;
Danger of collapse;
Removal of ground floors.

Action: Existing roofs are removed so that a concrete ring-beam can be inserted on external wall tops with the intent to strengthen the structure and hold the building together.
Negative Result: Loss, or part loss, of original roof;
Eventual horizontal cracking between rigid beam and underlying flexible structure.

Action: Treatment of perceived rising damp with chemicals.
Negative Result: The real problem is often one of the following: no gutters, leaking gutters, defective down-pipes, or ground level higher than internal floors;
Sand and cement plastering to internal walls is often part of the process.

So far, a great deal of unnecessary expense has been incurred, the building is on its way to being completely destroyed and it is not yet habitable. Here are some steps that should have been taken before the work began:

Expert Advice

- Seeking good, sound advice will save the building, its surroundings, money and time.
- Engage the services of an architect or engineer who specialises in architectural conservation before any work commences.
- An archaeologist may be required.
- Consult with the local planning authority and its conservation officer regarding the necessary permission and the nature of the work intended. There may be grants available.
- Through your architect or engineer employ a contractor and sub-contractors with experience to carry out the work properly.
- Consultants with specialist knowledge regarding damp, pointing, plastering, etc, may need to be brought in.

Research

- The building, its interior, surroundings, attendant structures and materials should be researched and analysed. Later decisions on what is to be done or not done will be based on these findings.

Minimum intervention

- Intervention should be made only to alleviate any further deterioration, for example, tying a structure together with stainless steel rods under floors to prevent collapse.
- Selective pointing where necessary rather than global pointing of the building.
- Repairs to flashings of chimneys, parapets, etc, to prevent water ingress and deterioration to ceilings in the building.

Like with Like

- Existing materials are repaired using matching materials, for example, mortars for pointing, plasters and renders replicate existing lime mortars.
- A wrought-iron gate is repaired using wrought iron and hot rivets rather than using steel and modern welding techniques.

Reversible intervention

- Any intervention made is capable of being undone, for example, an internal partition is constructed to divide a large room in such a way as not to damage the existing ceiling, cornice, picture and dado rail, or skirting. In the future the partition can be removed without causing damage.
- A door or window opening is blocked up without damaging the existing stone, or brick reveals, or overhead arch, or lintel, or sill.
- A soft-mortar repair is made to damaged stone.

Repair rather than replace

- Example: up-and-down sash windows are repaired using minimum intervention techniques rather than being replaced.
- Plasterwork is patched using like-with-like techniques rather than being replaced.

Finding the right people to do the work

Previous experience in modern construction does not automatically qualify any person to work in the area of architectural conservation. Without additional education and experience it could in fact lead to many problems. This can be explained when we realise that the technology we had up to quite recently (50 to 150 years ago, depending on where you live) was primarily Roman and had persisted in Europe and beyond for about 2,000 years. Most architects, engineers and craftspeople have little or no knowledge of traditional-type structures. This is slowly changing with third-level institutions offering degree courses in architectural conservation and professional institutions and contractors' organisations offering lists of their members who specialise in this area. For craftspeople, workshop opportunities are available to develop their existing skills in their relevant crafts.

The need for education at all levels can be summarised by the following chart, which shows the differences, in general, between the old and new technologies:

Parallel Technologies

Old building technology	New building technology
Solid wall	Cavity wall
Thick wall (600mm plus)	Thin wall (100mm)
Structural load-bearing walls	Non-structural walls
Lime mortars, renders and plasters	Cement mortars, renders and plasters
Flexible structures	Rigid structures
No allowance necessary for movement	Joints to allow movement
Permeable structures	Impermeable structures
Foundations sometimes non-existent	Concrete foundations
Arch or lintel	Concrete beam or steel beam
Permeable washes & distempers	Impermeable paints
No damp-proof membranes	Damp-proof membranes
Empirical design (rule of thumb)	Scientific design
Local indigenous materials (mostly)	Global materials
Few materials	Diverse/multitudinous materials
Intimate knowledge of materials	Little knowledge of materials used
Handmade	Machine-made
Made on-site (mostly)	Made off-site (mostly)
Long-life structures	Short-life structures
Low energy input	High energy input

Both technologies clash not only when old buildings are repaired or added to with modern materials but also when endeavours are made to incorporate older materials and techniques in new buildings.

THE SINGLE-STOREY VERNACULAR HOUSE

For many years, only major monuments were protected and restored and then without reference to their surroundings. More recently it was revealed that, if the surroundings are impaired, even those monuments can lose much of their character.

Today, it is recognised that entire groups of buildings, even if they do not include any example of outstanding merit, may have an atmosphere that gives them the quality of works of art, welding different periods and styles into a harmonious whole. Such groups should also be preserved.

The architectural heritage is an expression of history and helps us to understand the relevance of the past to contemporary life.

European Charter of the Architectural Heritage,
Council of Europe, Amsterdam 1975

These are the buildings disappearing from the landscape, the simple, vernacular buildings that draw little or no attention. Sometimes they stand alone as a farmhouse, maybe with a stone barn, at other times they lie in groups down a side street in a town, or as part of a deserted booleying village in the hills. I would suggest that these are the buildings we should try to preserve, and here we look at the elements that make them special.

Commonly, the individual farmhouse is neglected and allowed to fall into disrepair or demolished while a new house is built along-side. The row of small cottages in a town is demolished to make way for social housing of a modern standard, because to convert the older houses has been estimated to cost more, and anyway they are considered to be a reminder of poverty and nothing special architecturally.

This line of thinking will eventually lead to nothing being left except buildings of national importance, and the subtle variation that once

Vernacular building in stone and thatch, County Donegal.

existed from one place to the next and from building to building will
be gone forever. Many small vernacular buildings have not so much
disappeared from the landscape as been incorporated into larger
modern dwellings.

Vernacular farmhouse of stone, slate and corrugated iron over thatch, North County Dublin.

Small vernacular building with typical Irish roof structure, County Donegal.

Vernacular buildings are defined as being built without the assistance of an architect and nearly always of locally found material. Often there is little evidence of professional craftspeople like stonemasons having been involved – work was done by the owners, with the help of relations or friends. These buildings, therefore, vary according to whatever material and skill was available, the local style of building, and the social standing of the owner. Natural materials like stone vary according to the geology of the area. Thatch may be of cereal crops or reeds from a river. Other materials like mud, turf (peat), wattle and daub were also used.

Mud walls are to be found where the better land exists while stone is associated often with poorer land, but stone walls and mud walls are often used in combination with each other.

Similarly, stone is bedded in lime mortar, laid dry or with mud, depending on location. Vernacular buildings are to be found everywhere in the world where there is human settlement. Irish and other European vernacular styles were exported to North America, Australia and elsewhere, particularly in the nineteenth century when there was mass emigration. Vernacular buildings in general are constructed today only in developing countries; in the developed world they no longer exist as a living tradition. The available stock of vernacular buildings in developed countries is therefore dwindling and becoming more precious year by year. A varied way of life that once existed is being evened out by modern styles which are peculiar to nowhere and have no boundaries.

In Ireland up to the 1950s vernacular buildings were still being built and repaired, and quicklime was often drawn from the local kiln to limewash them once a year. The lack of variation in building styles and materials within a particular area over a long period of time makes vernacular buildings difficult to date. A tradition of building and repair passed down from one generation to the next and stretching back many centuries has come to an end in just a couple of generations. All of this, combined with lack of education and training in the understanding of their technologies, means that these buildings are now at their most vulnerable. We live in a transitional period when we have a duty of care to preserve them; they need to be repaired and adapted to modern living with sensitivity. In another generation or two they will, we hope, be more widely valued, understood, and maintained than they are at present.

The numbers of single-storey one-roomed vernacular dwellings that existed in pre-famine Ireland must have been vast. Many of them were not houses, but hovels and cabins. The traditional 'long house' or linear house in which people and animals lived together is an ancient structure associated with animal husbandry. In these buildings the fire of turf was lit on the floor next to a gable wall and cooking pots hung from projecting stones in the gable over the fire. The smoke from the fire would find its way out through the thatched roof as there was no chimney. Two doors opposite each other in the side walls facilitated cows being brought in for milking – through one and out the other. At night all animals would be brought inside; it was considered bad luck if the animals could not see the light of the fire.

From spring to autumn whole families near mountainous areas would take part in booleying – taking cattle to the mountains for summer pasturing which included butter-making. The booley houses would be even cruder, being constructed of sods of clay and grass or similar materials. Even so, it appears that booleying was very much enjoyed and was a time for social gatherings, crafts and singing.

Famine, emigration, the 'clearances' or removal of people and their dwellings off the land, and the natural tendency of materials like thatch and mud to disappear back into the landscape, combined with the later shame of poverty, means that relatively few of these dwellings are still in existence today.

Within the broad description of the single-storey vernacular house, there were different standards of comfort depending on social position, ranging from the simplest cabin built of dry stone, with a 'lean to' roof in sods, without windows or smoke-hole, to a small, thatched, mortared stone cottage with a smoke-hole or simple chimney, one or two small windows and maybe even a partition wall to create two rooms. In addition there may have been out-buildings. Any of the above could date from the seventeenth century to the twentieth century. From the eighteenth century onwards improvements began to occur slowly and gain momentum, in particular after the famine of the 1840s.

Although these buildings are the postcard- or chocolate-box image of Ireland which receives the most promotion, they are the least studied, understood or cared for building in the country. Generally quite small, they do not lend themselves to today's spacious living.

We will now analyse the details of the typical Irish vernacular cottage that many people alive today were reared in. They can be roughly divided into two types, direct-entry and lobby-entry.

Direct-entry houses

Main feature: Direct entry from the outside into the living area with the hearth on the inside of a gable wall. A rear door was often provided on the side wall opposite the front door.

A bed-outshot was common near the hearth. This was an alcove in the external wall projecting to the outside in which a bed could be stored (north-west Ireland, also Scotland and Scandinavia).

History: Originated from the long house, a time when humans and animals lived together in a single-roomed structure.

Width of house: One room.

Number of rooms: Originally one, but later divided or extended to provide two or three.

Location: The west and mountainous areas of Ireland.

Wall structure: Stone with two gables built dry or with mud/lime/sand mortar. Later rendered with sand and lime. Commonly limewashed over render, but also limewashed directly over the stone.

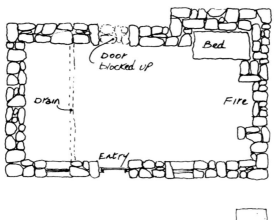

Direct-entry house (mountains).

Roof: Simple 'A' style structure, generally thatched, but later slated or sheeted with galvanised iron. Thatch, or the remains of thatch, is often seen under the galvanised iron.

Fireplace: On one of the gables. Originally the hearth would have been on the floor in front of the gable with smoke escaping through a hole in the thatch. Later (*c*. nineteenth century) fireplaces with a breast and chimney were built.

Lobby-entry houses

Main feature: A lobby created by a baffle wall opposite the entry. A spy hole was provided in the baffle wall so a visitor could be seen entering the house from the fireplace on the dividing wall at right angles to the baffle.

History: These houses were not used to house animals.

Location: Low-lying areas and in the east of the country.

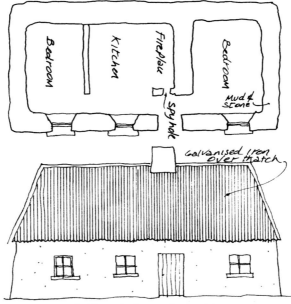

Lobby-entry house (lowlands).

Wall structure: Stone, mud or often both. Where stone and mud are used together for walling, the stone is used at the base of the structure.

Roof: Both ends hipped or half-hipped in thatch. Slate, or later galvanised iron, often laid over the thatch.

Fireplace: Central on the internal dividing wall.

Number of rooms: Commonly two or three. The internal dividing wall with fireplace may be built in mud and the hood or canopy over the fireplace in wattle and daub. Other walls are often only partition walls.

The limited amount of material used in the construction of any of these buildings is striking, often as little as five or six different types, such as stone, mortar, thatch, rough timbers and iron (latch, hinge and fire crane).

In a general sense some of the typical features of both houses are:

1. House only one room in width, approximately 12 to 15 feet wide.

2. Solid wall construction in either stone or mud, or a combination of

these. Usually very little cutting or shaping to be seen on stone. Sometimes surrounds to windows and doors in brick.

3. Mortars of lime and coarse sand, mud, or a combination of lime and mud. Also built of dry stone, and rendered.

4. External renders, usually wet dash (lime-based) and limewashed, traditionally white with strong colours on windows and doors, including reveals. Boundary walls may be built dry and limewashed. Mortared walls and circular piers in some parts of the country, again limewashed, often over a wet dash render in lime and sand.

5. Small sash windows and doors, sheeted, ledged and braced, sometimes as a half-door to allow light in, to keep animals out, and as a place to converse with neighbours. Timber lintels over doors and windows particularly on the inside, but also externally at times.

6. Single-entrance step or threshold in stone, or floor almost level with outside ground level.

7. Roofing material of thatch (wheaten or oaten straw, reeds, flax, marram grass, rushes – depending on locality). The thatch was usually fixed with scallops (hazel rods) to scraw (clay and grass sod). Other materials, like corrugated iron, eventually began to take over from thatch. Slate was also used and stone flags (depending on location) – often a corrugated iron roof was fixed over an old thatched roof. The iron roofs invariably were painted with a rust-resistant paint, red in colour. No gutters or down-pipes, which is understandable on thatch, but also often lacking on slate and corrugated iron roofs as well, resulting in dampness at the base of walls. Roofing timbers could be very rough and unshaped, sometimes just branches from a tree occasionally shaped with an axe and fixed with wooden pegs rather than nails.

8. Fireplaces sometimes have a wattle-and-daub hood supported on a wooden beam over the fire. The fireplaces could be 2 metres wide or more and either on a gable wall or an internal cross wall.

9. Ceilings might be lath and plasterwork or wood sheeting – or be non-existent.

THE CONSERVATION AND REPAIR OF VERNACULAR BUILDINGS

Later in this book, lime renders, plasters and washes are explained in detail. Similarly, aspects of working with stone are discussed. This chapter therefore concentrates on areas not covered elsewhere, such as roofing.

Newly thatched vernacular cottage, County Kildare.

Stone-flagged roof, County Clare.

Traditional farm buildings with galvanised roofs, circular piers and iron gate, County Meath.

ROOFS

A sound roof is probably the most important element of any habitable building. Roofs are one of the first areas to deteriorate rapidly in unoccupied buildings: thatch decays, slates slip off and corrugated iron develops holes from rust. The consequences are ingress of rain and dry rot of roof timbers, ceiling joists, plaster laths, timber ceilings, partitions and skirting. Very quickly holes appear in ceilings, and timber floors become unsound. Windows get vandalised and pigeons find a home to roost. A small vernacular house that has stood the test of time for maybe two hundred years and more can, within one year of neglect, develop serious problems. Many new owners are confronted with an expensive situation because they have arrived a year or more too late and are now faced with extensive repairs.

The repair of roofs is not the subject of this book but the following may be of help.

Simple three-bay vernacular house, which imitates the grand classical style, County Kilkenny.

Roof timbers

Traditional Irish roofs deteriorate mostly at the wall plate and rafter ends. A common building method from one end of the country to the other is to take the solid external wall up to the slope of the rafter, to lay the wall plate on the inside, and to slate or thatch over the top of the wall. To accommodate this, but not always, dummy rafters, often of a smaller size, are laid directly on the wall and bedded into position with mortar and stone. Slating battens are then fixed and the top of the wall slated over. Dry rot commonly occurs in these dummy-rafters and the wall plates. Repair and replacement of badly affected parts may be required, but not generally of all the roof timbers. Dry rot ceases to be active and dies when water is eliminated, and it does not spread outside wet areas. By supporting the existing roof internally it is possible to replace ends of rafters one at a time by splicing on new ends.

Collar ties are sometimes missing, having been used for other purposes, and nails fixing collar ties to rafters are often badly rusted.

The loss of collar ties will make the bottom of the rafters spread and push out walls. The sagging of rafters also is a result of inadequate collar ties. When refixing these ties, instead of nailing, which causes undue vibration to slated roofs, refix by screwing.

Slates

See Chapter 4, Understanding the Process, for details on slate sizes and fixing. Slates do not have an indefinite life span. Some are in good condition after two hundred years, others fail in a very short period. Slates can de-laminate and their iron nail fixings can fail. Second-hand slates are not always a good buy and some imported slates may vary in quality. Just like other areas of the house, slates do not all deteriorate equally and therefore selective replacement may be all that is required.

CHAPTER 4

UNDERSTANDING THE PROCESS

THE BUILDING OF A TWO-STOREY EIGHTEENTH- /NINETEENTH-CENTURY HOUSE

By way of illustration we will travel back in time and look at the procedures and materials involved in building a simple two-storey, eighteenth- or nineteenth-century house in stone. This will allow us to see what happened originally, so that we have a better understanding when it comes to analysing problems of deterioration and repair – and tackling the conservation of a building ourselves. These traditional houses are still plentiful and are rather plain in appearance with little decoration, but in their simple classical proportions they have a beauty that sits comfortably with the landscape.

This house may not have been designed by an architect, but by a local builder using locally available materials and labour. The clients – reasonably well-to-do farmers, clergy or business people from the local area or town – had a strong say in the design and layout. No two buildings were exactly the same and they usually comprised elements of both vernacular and classical. At ground-floor level, they had a central front entrance and fanlight with a window each side, while at first-floor level three windows made them three bays wide. Centralised chimneys, often two, or gabled chimneys were the norm. A basement was not unusual in the larger versions.

A two-storey semi-formal classical town house, Carrick-on-Shannon, County Leitrim.

Once the design and price was agreed the builder would carry out the following procedures.

Preliminaries

Lime and sand was soured out in advance for mortars, renders and plasters. The quicklime was drawn from a nearby kiln and some of it run to putty into a pit on-site for finished plasterwork. The remainder was mixed directly with sand and water to produce a hot lime mix, which was then soured out for mortars to build stone and brick, as well as for base coats in renders and plasters.

Cut stone elements such as quoins, sills and steps would be ordered from the local stonecutter. At times these are not from a stone indigenous to the immediate area, indicating that they were quarried, and possibly cut, a considerable distance away. Stonecutters often produced such elements as stock items. Also, itinerant stonecutters, like other craftspeople, walked the countryside looking for work and were prepared to set up temporary residence on-site while the work lasted.

Clay bricks were ordered for flues and for window- and door-openings. These were often burnt locally in clamp-fired kilns, using the green sun-dried bricks to build the clamp itself, and fired using

Brick clamp (turf).   Clamp (fired brick).

turf, wood or coal, whichever was available locally and cheapest. Sometimes on larger houses, itinerant brick burners made the bricks on-site from suitable worked sub-soils which were thrown into a moulding box, air dried, and then fired. Clamp-fired bricks varied in quality depending on how much heat they got in the kiln and the quality of the clay. At times the fuel was mixed in with the clay, so that the bricks were burnt rather than baked. A 'peacock's tail' or blue flame rising from the clamp after a number of days' and nights' burning indicated that the bricks were burnt. The quality of these bricks varied from being hard-burnt near the fire to being soft and porous on the outside of the clamp. The softer bricks may have been put back into the next firing or used internally where they would not be affected by the weather. The brick-clamps fired by turf required sods of turf (peat) to be hurled into the brick arches, as the tunnels so formed in the clamps were called. Mud was used to coat the exterior of the clamp and sheets of canvas and, in later times, galvanised iron were used to prevent the wind racing the fire in the clamp. A hard life, but one in which women and children were commonly employed.

Local rubble stone was quarried and delivered to site. Rarely was it transported any great distance, in some cases being quarried on the site itself. The major cost of rubble stone since medieval times and even today is transport, so to have a quarry nearby was a bonus.

Slates, some of Irish origin and many imported from Wales (particularly from the eighteenth century onwards) were also ordered.

Windows, shutter panelling, doors were made on-site or ordered from a joiner.

The staircase in these houses was a feature but was still relatively simple and not on the grand scale of the larger classical house.

Well-dried timber, some of which was imported, was used. The whole staircase with ballusters and curved handrail might be made on-site or partly off-site. Pine timber was most commonly used, while the handrail might be in mahogany.

Roof and floor timbers, slate battens, plastering laths, etc. were all ordered, often imported. The builder might have his own saw-mills and cut his own timber. Plaster laths for partitions were bought in bundles.

Setting out

The building was next set out on-site using large timber squares constructed with sides having a 3:4:5 ratio, which automatically produces a right angle of 90 degrees. Levelling was carried out using a plumb-rule attached to a horizontal sighting board by timber braces. Measuring-tapes and string lines were also used. Diagonals were checked to see if equal on right-angled areas.

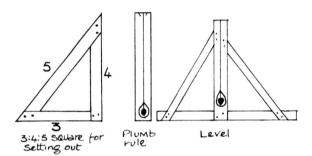

3:4:5 square for setting out   Plumb rule   Level

Sub-structure

The house under consideration has no basement but if it had then the site would be excavated to the floor of the basement which could be fully or partly underground. Basements often had windows for light and ventilation. Soil was either backed up against the wall or kept back by a retaining wall with a cavity around parts of the house to reduce dampness.

■ **The foundations** were dug to reach solid loadbearing soils – the agrarian society of the time knew their soils. Quite often little or no foundation was built: the width of the wall was sufficient to spread the load of the building.

- **Rising walls** (between foundation and ground level) were built, striking a level line on which quoin stones were laid at the corners, and the house was checked dimensionally and for squareness before proceeding above ground level. Non-hydraulic mortars, if used below ground level, had to be left exposed without backfilling of earth during the construction of the house to allow carbonation to occur. Hydraulic lime mortars were, of course, far preferable at this level allowing backfilling to occur soon after laying.

Corner of building with quoins.

Superstructure

- **Quoin stones** were often around 10in to 12in in height but could be bigger. If a stock item, they would be cut to the bed heights of brick so that, for instance, 10$\frac{1}{8}$in was suitable for three courses of brick 3in high with $\frac{3}{8}$in beds, or would suit three courses of brick each 2$\frac{5}{8}$in high with $\frac{3}{4}$in beds. Mortar beds increase with the irregularity of the bricks used.

  But the house we are considering was not built of brick except in certain areas so the quoin stone height did not have to suit brick sizes. The quoins in classical work are of equal height, width and length.

  It was not uncommon for the bed shape of the quoin stones to be nearly triangular and, although this was not good practice, it was done to save weight in transport and lifting and seems to have had little adverse effect on most buildings.

  The quoin stones, as stated earlier, were often imported from another district, so we sometimes see granite quoins in a limestone district.

  Quoin stones on rubble-faced buildings usually projected to allow for rendering, and were sometimes chamfered. An average projection would be 25mm. They are remarkable for the accuracy of their cutting and laying, usually with thin bed joints. One can imagine the oldest and most skilled stonemason carefully laying and plumbing the quoins.

I have heard accounts of quoin stones being used as a marker for a day's production so that the target each day was to rise one quoin stone on the external walls, excluding the internal walls.

- Rubble stone was then laid between the quoins in hot lime mortars. An average wall thickness was about 2 feet (600mm) but this varied from 18in (450mm) upwards for external walls. The thickness of the wall depended on a number of factors which included the following:

    a)   The size of the stone, so that a wall could be comfortably built with two faces without excessive cutting on the bed width

    b)   A locally-known width of wall that was known to resist rain penetration

    c)   The width-to-height ratio of a wall for a house was rarely a consideration

    d)   Basement walls were thicker and then reduced in thickness near ground level. Sometimes walls were reduced at each floor level internally, gables too in roof spaces

    e)   In pre-eighteenth-century work the width of the wall was often excessive, for defensive purposes.

- Damp-proof courses appear only in the later nineteenth century, the ideal place for them being just above ground level and below the timber floor.

- External faces of rubble walls were sometimes built with larger and better quality stones than internal faces. This was to reduce the number of mortar joints and to have a more weather-resistant stone facing the elements. At times, individual stones were tilted outwards to throw off water and also to create a small ledge on which to hang renders.

- External doors were met with as soon as rubble work was run between the quoins above ground level. On formal classical houses, cut-stone elements at entrance doors vary from the extremely simple to columns and entablature as part of a portico. Our semi-formal house has none of these features.

- Beneath windows from ground-floor to sill, walls were commonly reduced to allow easy access to the window and shutters inside. Brick was often used for this purpose as it is hard to build a 9in

(225mm) thick wall with rubble stone.

- Window sills were not always laid at this stage but later on. The practice was not to bed sills solidly under their centres, but only to point them externally. This was to prevent sills breaking their backs when either or both sides of the window opening settled in time. No damp-proof course was laid as happens today underneath and up the back of the sill. Sills were throated underneath to throw rainwater clear of the wall. Where this was not done problems ensued: walls stayed wet longer and dry rot, crumbling plaster and peeling paint occurred internally. On un-rendered limestone buildings today, the areas under effective window sills are often the dirtiest because they are not regularly washed down by rain, and there is a carbon build-up as a result. The same occurs under any other projecting architectural elements that shed rain effectively. Limewashing is prone to deterioration and streaking under non-effective projecting elements without drips.

- The window opes were next set out. Brick was commonly used on rubble-stone buildings which were later rendered. Brick allowed a rebated reveal to be built to accept the box element of the up-and-down sash window, while showing a 4½in face to the reveal. Stretching bond was used to accomplish this, usually 'block bonded' to three or four courses high. On more formal classical houses cut stone was used and left visible while on other houses, particularly of the later nineteenth century, the brick was built as a 9in (225mm) reveal but projecting and visible in a higher quality brick than considered in our example. The brick bonds used to accomplish this were usually Flemish or English bond.

- The rubble wall continued to rise between the external corner quoins, with the door and window reveals all being built together at the same time. Internally these reveals were nearly always widened to allow light in to the building which thick walls would exclude. The internal reveals were also often built in brick for ease of building, because of the cutting and shaping involved at corners which were not square. Horizontal timbers were built-in to fix shutter panelling later.

- Camber arches were the most common style over windows in the eighteenth/nineteenth centuries. The advantage was that they had little or no rise (3mm per 300mm of span = 1%), so that

door and windows did not have to be specially shaped on their
heads to a curve as with a segmental arch, which involved extra
expense. The rise also counterbalanced the optical illusion of
sagging given by straight horizontal lines. It was, of course, also

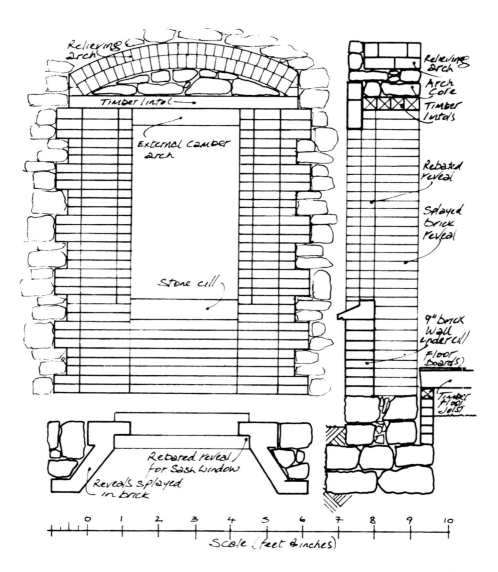

Elevation, section and plan of typical eighteenth- and nineteenth-century Irish window detail.

the style of much of this period; all brick houses in Georgian Dublin had camber arches, some cut with great accuracy. The back of the head of the arch formed a rebate, as in the vertical reveals, to accept the window.

- A semi-circular arch was built over the main entrance door-opening instead of a camber arch, so that a fanlight could be installed to add decoration and provide light to the inside hall-way. Elliptical and segmental arches were also used for this purpose. A semi-circular arch over a tall window was also common at the rear of the house to light the stairs.

- Timber lintels were employed internally over windows. A number of timber lintels, often much under-sized, were used to bridge the thickness of the stone wall between the internal face of the wall and the back of the brick camber arch. A relieving arch (brick) was built over the top of these lintels, on a solid core of brick or stone, to support the overhead weight and distribute it each side of the lintels down the jambs of the wall. The timber lintels were 'nailed' on their faces with projecting nails to fix internal plaster later.

- Putlog scaffolding supported the stonemasons and their materials. This type of scaffold, which was built of timber poles, utilised the building to carry the weight of the men and materials on the working platforms. To accomplish this, putlogs or horizontal tim-

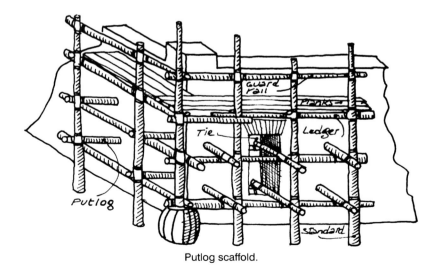

Putlog scaffold.

bers running across the width of the scaffold extended from the ledger to the wall. Standards or uprights were buried in the ground or fixed in large barrels filled with sand or soil. The timbers were lashed to each other with tarred rope (in later times wire rope).

The reason I mention this is that the holes, called putlog holes, in which the putlogs sat are sometimes noticeable and poorly filled. The heights between these holes tell us that a common lift height was about 4 feet 6 inches (1350mm), which was a comfortable height for a man to reach when building with stone. Putlog holes in more ancient buildings are considered very precious and should not be filled in, but left open with their bottoms sloped off with a hydraulic mortar. Old putlog timbers may be found in these holes and these can be dated. Putlog holes in older buildings were possibly always left open to accommodate the easy erection of scaffolding for maintenance.

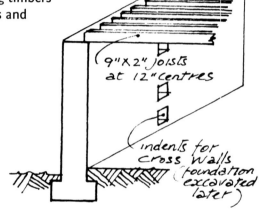

First-floor joists.

- Timber joists were laid just above the camber arches at 12in (300mm) centres for the first floor. Joists were commonly 2in thick and a traditional 'rule of thumb' to find the depth of the joist was ½ span + 2. So a span of 14 feet would be 7 + 2 = 9in. These were then used to support scaffolding on the inside of the wall but also added stability to the structure, sometimes forgotten nowadays when joists are removed and old buildings collapse. No flooring was laid at this stage, of course, just scaffolding planks. Joists were supported by vertical props underneath, so that excessive deflection or failure of the joists would not be caused by heavy material like stone.

- Internal walls at ground-floor level were started before external walls continued to roof level. The internal walls should have been constructed at the same time as the external walls with the

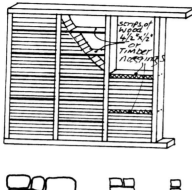

Top: 4½in brick nogged partition.
Above: Stone and brick partitions.

whole building rising at the same time. This stops differential settlement occurring where the ground is overloaded in one area compared to another, and also allows internal walls to be bonded into external walls, which cannot be done as well later. In practice, the internal walls were nearly always built later to allow access along the inside of the external wall for men and materials. Indents on the inside face of the external wall were left for bonding in internal cross walls, but sometimes forgotten. This led to vertical cracking at this connection later on, and sometimes the moving out of the external wall to such a degree that a modern intervention entails the use of tie bars.

Internal walls at ground-floor level were usually loadbearing and, therefore, built in either stone or brick. Stone walls, because of their width, take up too much space as internal walls and were therefore sometimes limited internally to cross walls with fireplaces where walls must be quite thick. Bricks were used elsewhere for 4½in or 9in walls. Loadbearing and non-loadbearing timber-stud partitions were often filled with 4½in thick brickwork (called brick noggins) which were laid stretcher-bond. The distance between the timber studs could be four bricks laid stretcher-bond (c.36in or 900mm), which was further apart than normal studs at c.12in. Horizontal strips of wood – 10mm to 12mm in thickness – called bonding slips were sometimes laid between horizontal courses of brick at intervals. These partitions were later plastered over, and it is a tribute to lime mortars to be able to cover such diverse materials as wood and brick.

Door openings in internal walls were formed slightly wider than the doorframe itself, unlike modern work where frames are built-in with the work.

- Fireplaces were incorporated in internal walls, in many houses on the two external gables. These fireplaces in gables are noticeable externally when they lose their renders, as the brick insertion in the rubble gable-wall can be seen winding its way up to the chimney. Very often this is where a vertical crack occurs for two reasons, expansion from the heat of the fire and because the weakest point in the gable was at the thinnest point of the wall. Also, differential settlement could break the back of the wall at this point. The house-type we are considering here has no fireplaces in the external walls.

Above the fire opening in each room a throating may have been formed with a sloped shelf over. The flues were constructed in brick which was much easier than with stone and more fire-resistant than limestone, in particular. The inside of the flue – c.9in x 9in (225mm x 225mm) – was parged (plastered) with

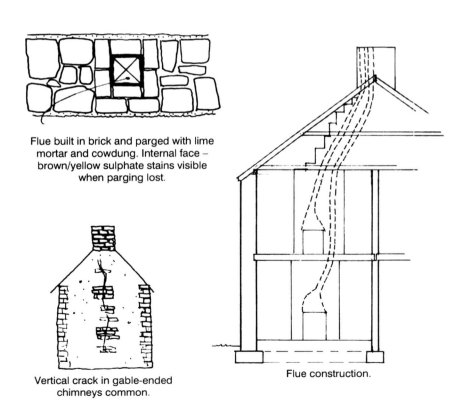

Flue built in brick and parged with lime mortar and cowdung. Internal face – brown/yellow sulphate stains visible when parging lost.

Vertical crack in gable-ended chimneys common.

Flue construction.

sand lime mortar (same mix as the flue was built with, ie, 1 lime : 3 sand). To this mortar cowdung was added, to reduce the effects of sulphates on the brick from the combination of flue gases, soot and water. Flues from different fireplaces at both ground floor and first floor were incorporated in firebreasts and brought up to first-floor level.

- Window openings were formed in external walls as before.
- External walls were then continued to roof-plate level all around the external perimeter of the house. The wall-plate was placed on the inside face of the external wall, and the top of the wall sloped down and outwards to a projecting flat stone on the external face of the wall. The roof-plate was bedded in lime mortar and levelled around the internal perimeter of the house.
- The internal walls at first-floor level were then built to ceiling joist level. More of these were likely to be of timber lath and plaster than found at ground-floor level and if so, may not have been built until the roof was slated unless they were partly load-bearing and supported the roof in some way.
- Fireplaces at first-floor level were constructed next and their flues incorporated in the same fire-breasts as those rising from ground-floor level. Very importantly all flues were kept separate. The firebreasts were then taken up through the roof space to emerge as chimneys with multiple flues. Flues were separated by 4½in 'withe' walls and sometimes capped, in later times anyway, with plain earthenware pots, yellow or red in colour.
- The roof timbers and ceiling joists were cut to size and fixed. Often in the simpler two-storey house collar ties were used at roof level instead of ceiling joists, resulting in a partly-sloped ceiling. Internal window-heads were then quite low and may have been sloped or tilted outwards to allow a view through the window while standing and to allow in extra light.
- The roof was battened to receive slates. The battens were some-times split in a similar fashion to those used for plasterwork, along the grain but thicker. As a result they were slightly uneven and twisted. In later times the battens were sawn. An average batten size was 2 inches x 1 inch. As stated earlier, slates were Irish or, more commonly, Welsh.

Before fixing, the slates were graded into thin (no. 3s), thick (no. 2s), and thicker (no. 1s). They were also confusingly called thirds,

Slating, using Welsh slates, Cork city.

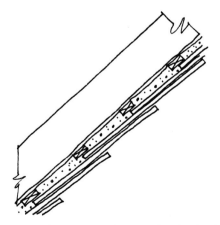

Rendering/parging with lime, sand and animal
hair to underside of slates.

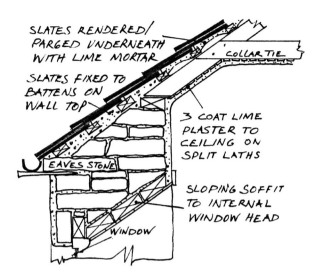

Typical detail on a vernacular two-storey house.

seconds and bests, which refers to their thickness and not their quality. The heaviest or thickest slates were fixed at eaves-level near the wall-plate, where the rafters can best carry the weight and where the greatest volume of rainwater would be. Medium-thick slates were laid next and finally the lightest or thinnest slates near the top of the roof. Slates that tapered in thickness along their length were laid always with their thickest end at the bottom. The slates were fixed with iron nails, unfortunately, which would later rust and fail. The lap was commonly 3 inches – this means that the third slate overlapped the first slate by 3 inches.

In order to 'hole' the slate a simple rule was used, still in use today. For example, with a Countess slate (20 inches x 10 inches), we take the length of the slate plus the lap divided by 2 plus ½ inch = (20+3)/2 + ½ inch = 12 inches. The slate is 'holed' 12 inches from its bottom edge. Next the gauge or spacing of the slating battens must be found by taking the length of the slate, subtracting the lap and dividing by 2, in other words, 20 inches – 3 inches ÷ 2 = 8½ inches. Below you will find a table of slate names and gauges.

SLATE NAME	SIZE (inches)	BATTEN GAUGE (inches) FOR 3 INCH LAP
Units	10 x 6	3½
Singles	10 x 8	3½
Doubles	12 x 10	4½
Headers	14 x 10	5½
Ladies	16 x 8	6½
Viscountess	18 x 9	7½
Countess	20 x 10	8½
Marchioness	22 x 11	9½
Duchess	24 x 12	10½
Princess	24 x 14	10½
Empress	26 x 16	11½
Imperials	30 x 24	13½
Rags	36 x 24	16½
Queens	36 x 24	16½

■ Two nails per slate were used and these were fixed about 1 inch to 1½ inches in from the edge of the slate. The nail holes were punched by hand from the back of the slate causing a counter sunk hole on the face of the slate in which the head of the nail sat. Slates were fixed about $\frac{1}{16}$ inch apart along their vertical edges. Very old roofs with thick stone slating were fixed with wooden pegs and not nails. Slates came in other 'in between' sizes also which had names, for instance 'Small Duchesss' at 22 inches x 12 inches and 'Wide Ladies' at 16 inches x 10 inches. On first-class roofs the rafters were boarded out first with rough boards, and then battened and counter-battened before slating. The counter-battening allowed air circulation to prevent timber decay. By far the most common practice in Ireland was to batten directly on to the rafters (no counter-battening), slate, and then render the underside of the slates with lime, sand (fine) and ox hair. Sometimes the hair was absent. In some southern counties in Ireland vertical slating of lime-rendered walls is still to be seen.

- Wet dashing (harling or roughcast) was applied to the external façade of the house using lime and sand. The work was started from the top and progressed downwards. One to three coats were applied. Putlogs were removed from walls in the downward journey and holes filled with stone and lime mortar. The late eighteenth and early nineteenth centuries saw the introduction of various patented cements which gave hydraulic sets. Roman cement from Britain was one of the better known with a very fast setting-time. Local hydraulic limes, when available, were also used. Portland cement began to be used from the mid-nineteenth century on. Renders were sometimes smooth and ruled to imitate ashlar stone.
- Windows (up-and-down sash) were fixed in window openings. Cast-iron or lead weights, slightly heavier than the sash were installed. Flax cords were used to hang the weights from pulleys within pulley-stiles which fitted in the rebated brick reveal of the window opening.

Elevation of sliding sash window.

- Floorboards were laid on joists. The ground-floor joists were laid on a timber plate which was set in mortar on a tassle or sleeper wall. In some cases when joists were laid their top edges did not align with each other because they had different depths. The tops of the joists were then adzed in position to achieve a flat plane surface over the whole floor area before the flooring boards were laid.

Stone-paved and earthenware-tiled floors were also common in certain areas of the house such as the kitchen.

- Doorframes (internal and external) were fixed.
- The main staircase was constructed minus the ballusters and handrail in case of damage, and the stair threads were protected.

Internal finishes

- Lathing of timber partitions and ceilings took place in advance of plastering. The laths were riven along the grain and although thin were flexible and strong. External walls, on their internal faces, were commonly studded and then lathed and plastered later for insulation purposes. Wainscotting of the lower half of external walls at ground-floor level was also common for the same purpose, and to hide dampness.

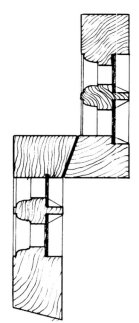

Section through both sashes (reduced in height).

- Internal plastering was carried out with mixes of lime, sand and ox hair. The lime putty from the pit on the site was used for the final coat (without hair). Stone, brick and lathed surfaces, including ceilings, were given a first or scratch coat, all with the same mix. See Chapter 14, Plastering. The surface was then scratched to receive the second or float coat of lime putty, sand and animal hair. The wall surface at this stage was quite flat. The surface was lightly scratched again to receive the finish coat of lime putty and fine sand.
- Cornices on ceilings were run in-situ in stages, using a running or horsed mould cut to the profile of the cornice. This was controlled by a timber batten called a running-rule fixed to the wall. Lime mortars were used to build out the profile shape and, for the finish coat, fine sand or marble dust with lime putty was applied. Decorative elements, such as acanthus leaves, could now be added to the cornice after being cast in plaster of Paris first and then fixed to the cornice.
  Elaborate decorative ceilings occur in the Great Houses and were

modelled in-situ from pattern books, using lime and a fine aggregate like marble dust.

- Skirting and picture rails were fixed to walls. Picture rails are nineteenth-century.
- Fireplace surrounds and hearths in marble, in cast iron (in the nineteenth century), were fixed in position.
- Distemper paint was applied to walls when plasterwork was dry. Distemper is made from whiting (chalk), animal glue, and pigments. Internal limewashing was confined to the poorer vernacular buildings, although basements and sculleries were often finished in limewash. Wallpapers were also extensively used in the eighteenth/nineteenth centuries. Oil paint was used to paint joinery such as windows, doors, skirting boards and so on. Limewash was applied to the external façade of the building over the wet dash, with the addition of an earth-pigment like venetian red.
- The gardens, during the course of the building of the house, would have been planted out and entrance piers, curved walls and iron gates installed, all in imitation of the Great House but on a much smaller scale. High garden walls (12 feet) were built, often in brick, to provide fruits early in the season for the dinner table, again in imitation of the Great House. An ice house might be a future possibility also.

The Great House deserves a mention in passing. The term palace seems more appropriate to some of these buildings than the word house. They are in direct contrast to the little single-storied vernacular house we began with. These large buildings were designed on the grand scale by well-known architects who specialised in this work and travelled widely.

Architects of this calibre had been to Italy and had studied the buildings of the renaissance, particularly those by Andrea Palladio at Vicenza and elsewhere. They were schooled in art, history, music, languages, drawing and painting, and knew the works of Vitruvius intimately. They had been to Rome and other Italian cities to study ancient Roman architecture. Their clients travelled extensively to Italy and to Greece in particular, as part of the obligatory Grand Tour.

The result at home was seen in imposing houses built to elaborate designs which, although beautiful, were in many ways more suited to sunnier and warmer climes.

In these large, formal, classical houses stone façades were meant to be seen, and to impress. The quality of work was high, often ashlar with fine joints, boasted faces and rusticated quoins. Pedestal, column, entablature and parapet rose in their glory and dominated the landscape. Everything conformed to classical proportion and detail, not only the building itself but also gates, piers and curved entrance walls into the demesne.

We know that Italian plasterers came to Ireland in the eighteenth century and, working from pattern books, created beautiful ceilings using lime putty and white marble dust.

The history of stonemasons is more obscure but, as it was and still is their custom to travel, we can assume that a certain number of European stonemasons were working in Ireland at that time.

Fine-jointed ashlar work became increasingly common from the eighteenth century onwards and there were few restrictions in the cutting, shaping or carving of stone.

GEOLOGY

If we understand how stone was formed and its natural constituents, then we are better able to understand how to use it and why problems may occur. Old stone buildings, in nearly all cases, reflect the geology of the immediate local landscape. There are, however, instances of stone being transported since the earliest of times, particularly over water.

But stone, being a heavy material, is not easily transported so having a quarry nearby was always of prime importance. Even so, on large medieval stone buildings, transport by horse and cart from nearby quarries accounted for much of the cost of building. A horse can pull about 1.5 tons of stone, which is only about 0.5 m³ of carboniferous limestone or granite. An average load by truck today is 20 tonnes (7.5 m³).

The quality of locally-available stone dictated the architectural detail possible in cut and carved stone elements. This is why at times stone for such details was imported from outside the area. The quality of stone available and how it was worked and laid could also determine whether a building needed to be rendered.

A study of old maps reveals that simple quarries and limekilns existed nearly everywhere. One of the problems today when we wish to repair traditional stone buildings is to find a suitable, available local stone. Often, the nearest working quarry is quite a distance away and does not produce a compatible stone. In no circumstances should local buildings, even when derelict, be dismantled and used as a quarry.

The knowledge of geology required for building and repair purposes is quite basic. The study of stone decay, biological and non-biological soiling mechanisms, and the assessment of cleaning methods such as water, abrasive, or chemical all demand a far greater understanding of geology than given here.

Stones throughout the world are classified into 3 main groups:

- Igneous
- Sedimentary
- Metamorphic.

IGNEOUS

Igneous rocks are formed from either magma or lava. Magma is a molten material below the earth's surface while lava is the same material occurring on the earth's surface (by volcanic activity).

> Magma ➡ plutonic ➡ below surface ➡ slow-cooling ➡ large visible grains ➡ good-quality dimension stone ➡ light colour ➡ granite, diorite and gabbro.

> Lava ➡ volcanic ➡ on surface ➡ fast-cooling ➡ small grains ➡ non-dimension stone ➡ dark colour ➡ basalt, andesite and rhyolite.

An in-between, medium-grained igneous rock is dolerite.

Granite

Granite is the chief igneous stone used for building. It contains the following minerals:

> *Quartz* (pure silica, harder than steel and extremely durable, commonly colourless). Sand is mostly composed of quartz.

> *Feldspar* (orthoclase = potassium feldspar and plagioclase = sodium feldspar). Feldspar is the dominant mineral in granite and responsible for its colour – white, grey, red, pink, green, etc. As hard as steel. When feldspar weathers it turns into a clay called kaolin which is responsible for decay in granites.

> *Mica* (black biotite and white muscovite). Laminated and noticeable when it is being worked as the chisel sometimes sticks between the leaves.

In addition to or in place of mica, hornblende and augite may be present.

Granites are generally found in mountainous areas. In Ireland granite was traditionally quarried and cut along the large batholith (large intrusive granite mass) which stretches from south County Dublin through Wicklow, Carlow and Wexford. It was also quarried and cut in County Down in the Mourne Mountains.

Granite is seen in stone buildings as a dimension stone (high quality, accurately cut) or as rubble. It was and still is popular for paving and kerbs because of its wearing qualities and its natural tendency to provide a slip-resistant surface. Of all the stones granite is probably the most respected, because of its hardness and durability. However, because of its hardness it wears out diamond-tipped saws and wires quicker than other stones. It is also severe on hand tools and in the past, before tungsten was used, granite would wear out tempered-steel punches and points in as little as twenty minutes. Granite is, of course, hard on those who work it by hand and can cause repetitive strain injury, particularly in elbows and wrists, from the impact when punching. Granite dust, like sandstone, is injurious to the respiratory system. For carving and sculpture it can be hard to read fine detail in granite, and it is best worked in large simple shapes. No natural beds are visible in granite as in many other stones, but the lie of the mica was used in the past to assist the splitting of blocks with wedges, and plugs and feathers.

SEDIMENTARY

Sedimentary stones used for building purposes are mostly limestones and sandstones. Both are softer than granite and are not as severe on tools. Limestone is geologically classed as rather soft but the harder carboniferous types and some sandstones are relatively hard to work by hand. Other stones, like oolitic limestone, can be cut with handsaws and worked with an axe.

Sedimentary stone is formed from the following series of actions:

pre-existing rocks ➡ broken down ➡ loose particulate material ➡ laid down in water or on land in beds ➡ compacted by weight ➡ cemented by calcite recrystallisation, iron oxide or silica ➡ resulting in limestones, sandstones, etc.

The pre-existing rocks from which sedimentary stone is formed as a sediment can be any or all of the igneous, sedimentary or metamorphic range. The breaking down of pre-existing rocks to produce sediment occurs due to weathering (water, wind, ice). This sediment may be a clay, sand or gravel. Sediment is transported by the action of rivers and wind and deposited in beds in sea water, lakes and deserts. Over time, bed is laid on bed and compaction occurs from the overhead mass. Cementing of these beds comes about by the action of calcite in the case of limestone and some sandstones, while iron oxide and silica also cause cementing in sandstones.

Limestone

Limestones are formed from sediment laid down in either seawater or fresh water. The most important mineral in limestones is their cementing agent, calcite.

Calcite (calcium carbonate, $CaCO_3$) is derived from shells and the skeletal bone structures of various marine animals. This is dissolved by the action of acidic water and redeposited among the sediment, causing a cementing action. Calcite is a soft mineral and dissolves with acid rain. Sulphur in city environments combines with calcite to form calcium sulphate which causes decay in limestones.

Irish limestones date from the carboniferous period and are well-compacted and relatively hard. Limestones were laid down in other periods as well.

Generally the older the stone the more compacted it is.

Oolitic limestones, such as Portland and Bath stone from England, belong to the Jurassic period and are therefore relatively young and easy to cut and shape. These were chemically deposited, which means that calcium carbonate was dissolved by acidic water containing carbon dioxide, and then redeposited amongst sediments. These sediments (grains of sand, small pieces of broken shell, etc.) were near a beach where wave action continually rolled the sediment back and forth until it became coated with calcium carbonate to form egg shapes. These were compacted in beds to form oolitic limestone, which was much used from the seventeenth century until the mid-twentieth century, mainly as a dimension stone for classical buildings.

Sandstones

Sandstone may be formed on land in wind-blown dunes in desert-like conditions or in fresh or salt water. The cementing agent classifies their description, for example

Siliceous sandstone – silica as a binder, which forms

durable sandstones, such as gritstones.

Calcareous sandstone – calcium carbonate as a binder produces sandstones which are susceptible to acid decay from polluted atmospheres.

Ferruginous sandstone – iron oxide as a binder deposited in solution. The iron oxide produces a variety of colours such as red, brown, brown/yellow, etc. Generally good weathering stones.

Argillaceous sandstone – mud as a binder producing poor weathering stones.

The durability of sandstone is determined by its binder, thus siliceous sandstones with quartz grains and a silica binder are the most durable.

METAMORPHIC STONES

Metamorphic stones derive from other stones that have gone through a change. This change is brought about as a result of

Heat (from the interior of the Earth) and/or

Pressure (from overhead mass and movement of the Earth's crust).

Metamorphism causes

Recrystallisation of existing minerals

Compaction, which increases density and reduces porosity.

Igneous, sedimentary and metamorphic rocks can all be metamorphosed as follows:

Limestone to marble

Shale to schist

Mudstone to slate

Sandstone to quartzite

Granite to gneiss.

For building purposes the most common metamorphic stones are:

Marble

Marble is limestone metamorphosed and is used for decorative purposes. In the past marble was used chiefly for fireplaces and floors. The Parthenon in Athens was built in the fifth century BC from

Pentelic marble quarried 16 kilometres away. The most famous marble in the world is Carrara marble from Northern Italy, used by Michelangelo in the sixteenth century. It is snow white and translucent, metamorphosed from pure limestone, and used extensively all over the world even today for fine carving and decoration.

Limestones, when polished, are sometimes referred to as marble which, of course, they are not.

Irish Connemara marble is green in colour and very hard.

Marbles do not weather well in the outdoors in Northern Europe and are therefore best used indoors.

Slate

Slate is mudstone metamorphosed to slate by lateral pressure and heat. The most famous slate in the world is probably Welsh slate, which was exported extensively to Ireland from the eighteenth century onwards. Welsh slate is commonly called Blue Bangor slate in Ireland, but many other quarries also existed in Wales. Bangor slates are from the Cambrian period; other Welsh slates are Ordovician and Silurian.

Ireland produced slates from quarries in Tipperary (Killaloe), Donegal, Kerry and Kilkenny.

Slate, traditionally, was split by hand tools and graded by thickness so that the heaviest slates were laid first at eaves-level on the roof, then the medium and finally the lightest slates at the top of the roof.

Quartzite

Quartzite is sandstone metamorphosed. It is capable of being split but very hard to square up on the face, except with special diamond-tipped cutting saws. In areas where it was indigenous it was traditionally used for building purposes without cutting and shaping. Today it is used for flooring.

CHAPTER 6

STONECUTTING

In working on stone buildings we need to be able to recognise what we are actually looking at in order to analyse it, and we need to have a grasp of the process of stoneworking in order to apply it. In *Irish Stone Walls* I covered two areas of stonecutting: the establishment of a face on rubble stone without any working of the beds, and the cutting of quoin stones which requires the cutting of faces, beds, and ends. We will briefly recap on the principles of doing this so that we can progress to more advanced work.

The cutting of a single face on rubble stone is based on taking the arrises or face edges 'out of twist'. A flat plane surface is said to be out of twist.

The following is more the stonemason's craft rather than the stonecutter's. The result is four arrises 'out of twist' on the face of a rubble stone.

Typical stonemason's/ stonecutter's tools.

Cutting rubble stone

1. Lay stone on the ground on its natural bed. This applies in particular to sedimentary stones, but also to metamorphic stones which have visible bedding planes.

2. Choose the face of the stone. This may not be the longest side, as it is good practice to run the length of a stone into the heart of a wall.

3. By visual observation choose the furthest point of damage, indent or depression back from the face of the stone.

4. Lay a flat steel bar on the top bed of the stone, with its forward edge in line or just slightly back from the point of damage.

5. Scribe the top bed to the straight edge.

6. Pitch just forward of the line leaving the line still visible when completed.

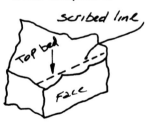
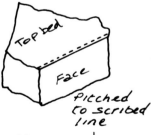

7. Turn the stone upside down, lining up the pitched arris with a line drawn or scribed on a concrete floor, or place a steel flat bar in tight to the pitched arris on the ground as shown.

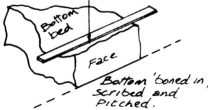

8. On the new top bed now place a second steel flat bar and sight or bone this in (ie, parallel the two straight edges by eye) with the one underneath. Scribe the new top bed and pitch as before. We now have a top and bottom arris in line with each other and 'out of twist'.

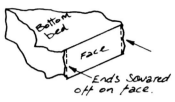

Ends Squared off on face.

9. Square off the ends on the face.

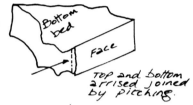

10. Pitch the ends to join the top and bottom arrises.

Top and bottom arrised joined by pitching.

11. Clear away any excess or projections along the ends which would interfere with placing the next stone tight against this one and thereby creating a wide vertical joint (a perp).

Ends cleared.

Finished Stone

A 'rock-type' face on a block of rubble stone 'out of twist' can be achieved in commercial works by the use of a hydraulic guillotine-machine in combination with a conveyor-belt. Rubble stone is fed on to the conveyor-belt by an operative, and transported by the conveyor-belt to the hydraulic guillotine where it is positioned in place under the blade of the guillotine. By compression, the force of the guillotine blade fractures the stone on a line, which achieves a rough face surface more or less out of twist. The ends need to be then squared up on-site by hand.

In more advanced faces, beds and ends are worked to a high level of accuracy, each square with the other. When this stone is laid the joint sizes are small and controlled. When the joint sizes are 3mm and less (sometimes 1mm) then this work is classified as ashlar.

From the beginning, stonecutting has been based on the skill of taking a surface 'out of twist'. Once this was achieved other surfaces could be squared and measured off this.

Traditional carpentry was based on the same principle of using an adze, wood or steel plane to achieve a flat surface on timber. This skill emerged in stoneworking thousands of years ago, and with it great buildings and even cities were built. Today, diamond-tipped saws can produce a flat plane surface in a

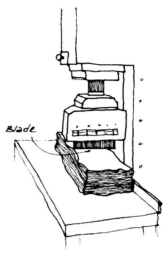

Hydraulic guillotine.

very short time, and also cut other surfaces at right angles to it. Even so, there is still a place for those who work by hand. To work by hand is to know intimately how earlier work was produced. This is most useful in the repair and conservation of old buildings. The skill of working by hand has no limitations, and design is not governed by the dictates of a machine. Those who work by machine should, therefore, first work by hand to realise the true potential of stone.

We will now have a look at the traditional skill of stonecutting, beginning with a quoin stone.

Derelict stonecutter's cottage, Ballyknockan, County Wicklow.

Quoin stone

Hoisting a stone window mullion into place, Drimnagh Castle, Dublin.

A stonecutter forging stonecutting tools (Italy).

A stonecutter working with compressed air tools.

Preliminaries

1. In this case we start with a bank. A stonecutter's bank is a stout table or bench. Alternatively, a large block of stone is used as a bank.

2. Stonecutters do not turn rocks into finely cut stones, they begin with stones which are quite close in size and shape to what they need. These stones, unlike many of the rubble-stone variety, are always of high quality and behave in a consistent manner when being cut. The quarryman's skill was to deliver a stone to the stonecutter's bank which was all of these things. Today, saws easily accomplish this task with dimension stone.

   The stone is banked securely so that it doesn't rock about or fall off the bank.

   It is worth explaining in brief the procedure to be outlined. The top bed of the stone is worked first, followed by squaring down off this bed to create the faces of the stone. The bottom bed is paralled off the top bed and worked together with the ends of the stone.

3. The first stage is to take the top bed 'out of twist'. Starting with the top bed, by visual observation (on larger stones by

a straight edge and dropped square) assess the lowest point.

4. The lowest point is marked on the vertical face of the stone. Now, with a straight edge held at the bottom of the lowest point, scribe a line roughly parallel to the top bed, making sure to stay below any secondary points of damage.

Scribe line below lowest point.

5. Pitch above this line leaving the line just visible.

6. With a 25mm wide chisel cut a draft on the top bed to the scribed line. This needs to be checked for accuracy with a steel straight edge. A rusted one is best as it will leave a rust stain on the highest points that still need to be worked.

Pitch and cut drafted margin.

7. Stand the straight edge on the drafted margin and go to the far side of the stone with a second straight edge.

8. Hold the second straight edge against the back face of the stone and sight against the first straight edge. Scribe a line on the back face of the stone. Instead of using a second straight edge it is possible to 'bone in' by eye instead.

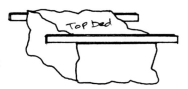

'Bone in' opposite side.

9. Pitch and draft to the second scribed line as before.

10. Now cut two more drafted margins at the ends, joining the first two on the face and back of the stone.

Cut second drafted margin.

11. The centre waste between the margins is now extracted by using a punch. This will

'Bone in' both drafted margins.

Cut last two drafted margins and remove centre waste, 'square up' bed on plan.

effectively reduce the waste down to within about 3mm of the drafted margins. On ashlar work which has 3mm mortar beds, the maximum tolerance in any one stone is 1.5mm above the drafted margins. It is not good practice to cut hollows in beds to achieve tight joints as this puts pressure on the external margins of the stone resulting in spalling.

12. We now have a top bed which is very nearly a flat plane surface. To reduce the waste further wide chisels, claws and bush hammers may be used, although these are usually reserved for the face of the stone and beds are usually left with a punched finish.

13. The bed is squared up on plan to the desired length and width of the stone by pitching vertically down the faces and ends.

The first stage is now complete.

The second stage is to establish the face, the ends, and the bottom bed.

14. Using a steel square, square from the top bed down the chosen face of the stone at both ends. Scribe and pitch. This establishes the length of the stone on the face.

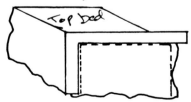

15. Again, square down at both ends of the stone but this time scribe and pitch on the ends. This squares the face of the stone with the bed.

Square down face, pitch and draft.

16. Measure from the top bed down the face and ends of the

stone and scribe a line parallel to the top arrises. This establishes the height of the stone.

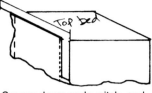

Square down ends, pitch and draft.

17. Turn the stone on its side and pitch this scribed line on the face and on both ends.

18. Now cut drafted margins at the top and bottom arrises of the face and bone in.

Parallel bottom bed from top and continue procedure.

19. Connect these top and bottom arrises on the face of the stone with two more drafted margins at the ends.

20. The excess waste on the face of the stone is first worked by a punch or point, then a claw chisel or bush hammer and finally a broad chisel. The finish required may be any of these.

21. The ends of the stone are now treated similarly.

22. Finally the bottom bed is worked by squaring from the face of the stone.

An alternative method used on large stones is to begin with four equally sized blocks of timber say 40 x 40 x 40 mm or 50 x 50 x50 mm. It is important that they are exactly the same size.

At each corner of the top bed of the stone cut out a small level table just big enough to seat each block of wood. These four tables are below any point of damage on the top bed. Two straight edges are now placed on top of each pair of wood blocks and sighted/boned in. In most cases it is only necessary to reduce one table until the straight edges sight perfectly. The four corners are now 'out of twist'. Lines are scribed between them and pitched before drafted margins are cut as above.

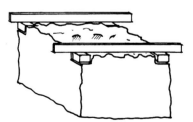

Taking 'out of twist' using hardwood blocks.

This is a useful method on large stones and on very hard stones.

On very large stones, it is hard to judge when removing the centre waste if you are above or below the drafted margins. Three-point boning is the solution where three objects (each the same height) are used, two on known, fixed points (opposite drafted margins), and the third on the lowest considered point of the waste in the centre. The eye can now sight from one known fixed point to the other, and the point in between is 'boned in' by reducing the stone surface with a punch or point.

A technique favoured by the Egyptians was to use three wooden sticks each of equal length. A string line was pulled between the two sticks on fixed, known points, and the third stick in the centre held against the line to show how much waste it was necessary to remove until it came level with the string line. They sometimes used this method also on wall faces which were already built, but were faced afterwards in-situ.

Cutting a chamfer

A simple chamfer is seen on many different cut stone elements in buildings, from the tops of plinths to string courses, door and window reveals and window sills and small chamfers on the arrises of quoin stones.

Chamfers prevent damage occurring to corners and direct rain off otherwise horizontal surfaces; they are decorative, also, and show interesting patterns of light and shadow on a building. Chamfers, as tangents, are used exclusively in cutting circular columns, bull noses, etc. and will be discussed later.

Now that we have conquered the taking of a face 'out of twist' and a second face at right angles to it, we are ready to move on to the cutting of chamfers.

With so much dimension stone readily available today this operation is often carried out on a stone which has been squared up

Cutting a granite window quoin with rebate and chamfers, County Kildare.

on the saw first.

The following are the necessary steps in cutting a chamfer by hand:

1. A chamfer is often, although not exclusively, cut at an angle of 45 degrees. A very common chamfer width on many buildings is 3 inches (75.2 mm). Cut a templet out of stiff plastic sheet to the shape of the chamfer required. The templet should be cut so that both its top edge and face edge are at right angles to each other, and can be held against the stone's top and face edges to achieve correct alignment. These lines can be double checked each end with a bevel off the top bed or face of the stone. Two lines are now scribed from one end of the stone to the other, along the top bed and face of the stone.

2. When working with highly-accurate dimension stone, it is sometimes useful to know how far to measure in from the arris on both the bed and the face to achieve a 45 degree chamfer of 3 inches.

   The chamfer represents the hypotenuse of a right angled triangle, and in the case of a 45 degree chamfer the distance to be measured in from the arris on both the top bed and the face is the same.

   The square on the hypotenuse is equal to the sum of the squares on the other two sides. In this case the hypotenuse or chamfer measures 3 inches, and this squared equals 9 square inches. Each of the other sides therefore measures $\sqrt{4.5} = 2.121$ inches (which for practical purposes is taken as 2.125in or 2⅛in).

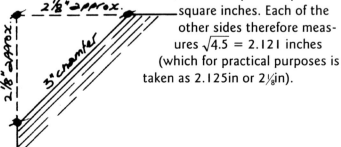

3. If the scribed line along the top bed is a reasonable distance in from the face then the waste can be pitched off. This is done very carefully, as it is easy to exceed the scribed line on the face. The controlling factors in this are the distance back from the arris at which the pitcher is applied, and the angle at which the pitcher is held.

   On small chamfers a chisel may be used for the pitching, and at other times no pitching at all is used, with the chamfer just worked longitudinally from one end of the stone to the other using a chisel.

4. After pitching, the waste is removed carefully with a point.

5. The chamfer is then worked and finished with chisels, sometimes with a claw chisel first, preferably finishing with a broad chisel connecting both scribed lines with a flat planed chamfer.

6. Chamfers are stopped or finished in a variety of ways. The styles of chamfer stops used in the past allow work to be dated.

chamfer stop

Curved work

To cut a stone with a curved face, such as seen on a bow-window, or on a curved stone entrance, the following procedure is followed:

1. The work is set out full scale to the appropriate radius using a trammel. In the case of very precise work like ashlar, the height of every stone is pre-determined and often the length as well.

   Templet (bed Mould)

   Curved work.

2. A templet is scribed, using the trammel as a guide for both the face curve and the two ends radiating towards the centre of the scribed curve. The templet is sufficiently long so as to accommodate any stone in the wall.

3. The templet is applied to the top and bottom beds of a stone which are 'out of twist' and parallel to each other. The face curve and the radiating ends are scribed.

4. The stone is pitched to the scribed lines and the waste removed with a point.

5. The face is dressed as required.

Columns

Columns begin as a length of stone with square ends. This is converted to an octagon by cutting chamfers, then to a sixteen-sided polygon – again by cutting chamfers, and finally to a circular column. The octagon is the favourite polygon to begin with because it fits within a square, requiring the cutting of only four chamfers to achieve an octagon. Stonecutters traditionally worked from a square to an octagon to a circle for practical reasons, but this took on a religious significance also, and in the case of many church towers and spires we can see a square tower worked to an octagonal spire which diminishes to a point, all reflecting the transition of our life from this world to the next.

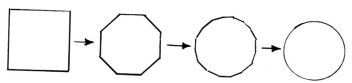

Larger columns are usually constructed from a series of drums, which are bedded one on the other to form the full length of the column. A column in true classical style has convex curved sides, and is also tapered. The convexity is called entasis and is used to overcome the optical illusion that a column is concave or waisted. To simplify the explanation of the process, the column drum we will consider here has the same diameter at either end and is neither tapered nor displays entasis.

1. Draw a square on a stiff piece of templet plastic with sides equal to the diameter of the proposed column drum. Within this square and touching its four sides draw a circle which represents exactly the required drum. Between the circle and the square draw an octagon touching both the circle and the square. Puncture a small hole through the templet at the centre point.

   AD QUADRATUM, a square revolved on its centre axis is a traditional method of setting out and most useful for developing circular shapes.

2. On a stone which is square in section to the same size as the templet and of the correct length, mark the centre point

at both ends by first drawing diagonals. A straight edge should be held horizontally at either end on the centre point, boned in by sight one with the other, and a horizontal line scribed at either end. This line is used to ensure that the applied templets

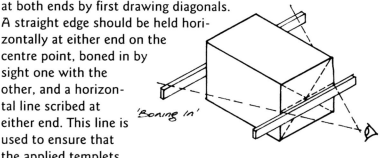

'Boning In'

are placed at either end in correct relation to each other, ensuring that all cuts from one end to the other are out of twist. With modern machinery the square block of stone now comes delivered perfectly true.

3. The plastic templet is now cut so as to reveal the octagon. The octagon is marked on both ends of the stone, using the horizontal scribed line on the ends of the stone as an alignment reference, with a similar line marked on the templet. The templet is also held with its centre point aligned on the centre point on each end of the stone.

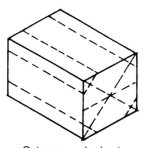

Octagon marked out.

4. Lines are scribed from one octagon to the other, these represent the four chamfers to be cut.

5. The chamfers are cut as described earlier, joining one octagon to the other. Straight edges are applied regularly along the chamfers to check for accuracy.

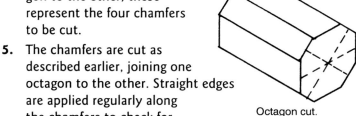

Octagon cut.

6. When the octagon is complete and accurate, the octagonal templet is applied to each end again with its centre point over the centre point marked on the stone. The templet is revolved until the eight corners of the octagon cut in the

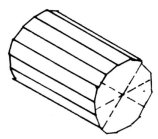

Sixteen-sided polygon cut.

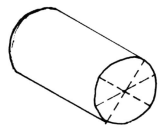

Finished column drum.

stone align with the centre of each of the eight sides of the octagonal templet. These sides are now scribed and cut to reveal a sixteen-sided polygon.

7. What is now left is very nearly a cylindrical drum, except for sixteen small ridges running the length of the drum. These are usually taken off by eye.

8. The column drum is now complete except at each end a square hole is cut to take a dowel which will assist in alignment when bedding and also prevent future mis-alignment.

Lathes have been used this century to produce columns. A circular diamond-tipped blade applied to a revolving block of stone very quickly produces drum shapes.

To produce a tapered column traditionally requires only that we apply a smaller octagon to begin with at one end of the stone, so that when the chamfers are cut they are tapered and reduce in width as they travel from one end of the stone to the other.

Classical columns with entasis or swelling are more difficult to cut because of the convex shapes, which means a straight edge cannot be run along a chamfer to check it for line. Instead, a special curved edge must be continually applied to the face of the stone to check the convexity of each chamfer as it is cut.

Balls

With the knowledge of how to cut a column by hand we can adapt this same skill to cutting a ball or sphere. These are to be seen on the tops of entrance piers to eighteenth- and nineteenth-century houses, sometimes obviously cut freehand without recourse to geometry, and looking a little odd as a consequence.

Begin with a cube of stone which will have six equal sides unless a base is required under the ball as an integral part.

In brief, a ball is cut by starting with the same exercise as before, the production of a cylinder. This is done starting with a square, then an octagon followed by a sixteen-sided polygon and then a circle. The same process is now applied to the cylinder, vertically, to create a ball.

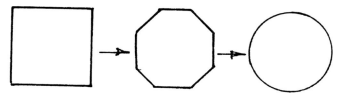

The stone is first worked on plan ...

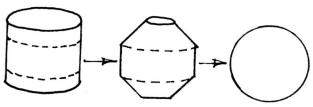

and then in elevation.

1. First cut a templet of a square, octagon and circle as before. Apply this to either end of the stone. The stone is then worked to a cylinder as before.

2. At either end of the cylinder from the centre point draw a circle with a dividers which has a diameter equal to one of the sides of the octagon.

3. Measure the difference between this newly-scribed circle and the circumference of the cylinder. Measure this same distance on the face of the cylinder from either end and circumscribe around the face of the cylinder.

4. Cut a chamfer either end of the cylinder using these two lines as guides. The width of this chamfer will be the same as cut earlier when cutting the octagon to produce the cylinder.

5. Now we have a vertical octagon, so next we must create a sixteen-sided polygon by first measuring either side of the sharp points of the octagon the same distance as we did to produce the cylinder. These lines are circumscribed around the stone and cut as chamfers.

6. The remaining work is completed by eye. A bush hammer or point is most useful for this work.

There are other ways of doing this, some quite crude, which result at times in humorous shapes.

Hollow, concave, or negative shapes are more difficult and are usually achieved by using templets or moulds which are applied to fixed running points to check the work as it proceeds.

The above skills, although strongly based on geometry, are as easily applied to free curves. Curves drawn by hand full scale on paper can be transferred to stone as tangents, which can then be cut as chamfers and finally worked down to smooth flowing curves. Classical Greek mouldings were freely drawn like this, while the Romans used the compass.

Eighteenth-century Custom House, Dublin.

CHAPTER 7

ARCHES

Romanesque

The first arches to appear in Ireland were not true arches but arch shapes cut in solid stone lintels. Appearing first somewhere between the seventh and tenth centuries in small Early Christian churches, they were simple, un-ornamented and Romanesque in style, being semi-circular. Arch shapes like these, cut in solid stone, persisted long afterwards and can be seen on castles built much later.

True arches made up of individual stones (called voussoirs) were seen first in the Early Christian period as relieving arches over stone lintels.

The first true ornamented arches in Romanesque style appeared in Ireland in the twelfth century. A short wonderfully inventive period followed when Romanesque architecture took on a distinctive Irish style. Within this style arches were semi-circular and the voussoirs cut from the softer stone varieties such as sandstone.

Early Christian (medieval).

Early Christian relieving arch.

The semi-circular arch was also used in vaulting of this period. By its nature the semi-circular arch causes horizontal thrust near its springing point. To resist this thrust, walls were built quite thick with only small window and door openings.

Gothic

By the end of the twelfth century Gothic influence had arrived and with it the pointed Gothic arch. These are still to be seen throughout Ireland on castles and monasteries, over doors and windows not built with multiple voussoirs but often as two single stones cut to a Gothic curve and meeting at the centre top in a point. The extrados, or outer edge of these arches, was invariably 'softened' or unworked, a feature also seen on quoins and door and window jambs where no unnecessary work was done.

Chamfers were common on the intrados to prevent accidental damage to the arris and also as a simple decorative device, sometimes finishing in a carved chamfer stop at the bottom of the jamb, which allows them to be dated.

Over the top of the extrados a projecting hood was sometimes used to provide shelter and decoration. These ended in a variety of styles called hood stops or label stops, some of which were carved.

The Gothic arch has less thrust than the semi-circular form and is widely seen in vaulting, using ribs to transmit weight to walls or directly to the ground, thereby freeing walls to be opened up as windows allowing light in.

Primitive vaulting from the fourteenth to the sixteenth centuries on castles in Ireland often shows the remains of wickerwork mats still underneath. These were obviously used as part of the centering or support system while the vault was being built. The mats could be laid over

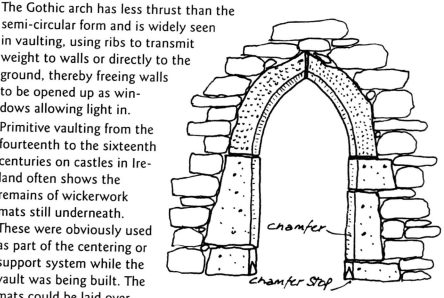

Gothic arch (medieval).

timber formwork only partly planked, as the mats were quite strong over short spans. Alternatively, solid centering built from stone, mortar and earth could be used. Mats would be placed over this to form the curve on which to build the vault.

ARCHES – SETTING OUT

The following points are general to all arches:

- All arches, except the true ellipse (which cannot be drawn with a compass), radiate from a centre point or points. These centre points vary from one to four depending on the style of the arch. The centre point in arch construction is called the striking-point.
- The base line of an arch, or the point from which an arch is built, is called the springing-line.
- The distance from the springing-point to the highest point on the underside of the arch is called the rise.
- The horizontal distance between springing-points is the span.
- The sloped sides of the wall, cut and radiated to the striking-point of some arches, act as an abutment and are called the skewback.
- Individual units (stones or bricks) in the arch are called voussoirs.
- The central voussoir at the highest point of the arch is called the keystone.
- The outside top of the arch is the crown.

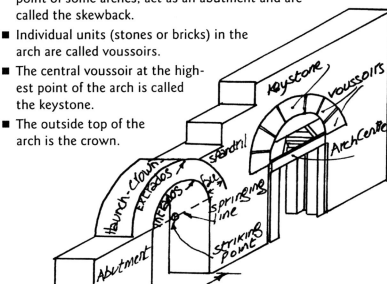

Arch terminology.

- The outside bottom sides are called the haunches.
- The bottom or internal curved face arris is called the intrados.
- The top or external curved face arris is called the extrados.
- The face of an arch is the vertical part seen, which is measured between the intrados and extrados arrises or lines; these are not always parallel so the face width can vary.
- The solid work between two arches is called the spandril.
- The solid work on either side of an individual arch is called an abutment.
- The arch support used for the construction of the arch is called the arch centre.

Semi-circular arches

Semi-circular arches are seen in a whole variety of places, from rough relieving arches reducing the weight on stone and timber lintels underneath, to ashlar cut work, carved and decorated.

The semi-circular arch considered here is of the simple, cut, undecorated type. If a chamfer is required see section on cutting a chamfer in Chapter 6, Stonecutting.

The work involved from the set-out to the cutting and building is discussed here in detail so that it will not have to be repeated on other arches, which, although different in shape, follow more or less the same approach.

The semi-circular arch is set out from a single centre point at mid-span. The rise is half the span which is equal to the radius.

The voussoirs radiate to the centre point of the arch. When the voussoirs are cut wedge-shaped their sides radiate to the centre point of the arch.

Semi-circular arch in porphyry stone, Lambay Island, Dublin.

The following is the procedure required to set out, cut and build a semi-circular arch.

1. Given the span and the face height of the voussoirs, set out the arch full-scale on a flat sheet of wood or on a timber floor or similar surface. This is best accomplished with a trammel, which can be as simple as a wooden lath with a nail fixed through to the wooden floor. From this point, measure the rise to give the intrados, and the face height to give the extrados. Radiate the trammel with a pencil held at these points to give two concentric semi-circles.

2. The arch is now divided into the required number of voussoirs. An uneven number will result in a central one at the point of the arch acting as a keystone.

   We will assume the arch has fifteen voussoirs which includes the keystone, a span of 2 metres, and a face width from intrados to extrados of 400mm. The diameter of the arch to the extrados line is 2 metres + 2 x 400mm = 2.8 metres. The joint width between voussoirs is 6mm.

   In practice, a pair of dividers is used to walk the extrados curve until it takes fifteen steps to travel from the springing point on one side of the arch to 6mm short of the springing line on the other. This takes less time than one might think.

   The reason for measuring or stepping out voussoirs on the

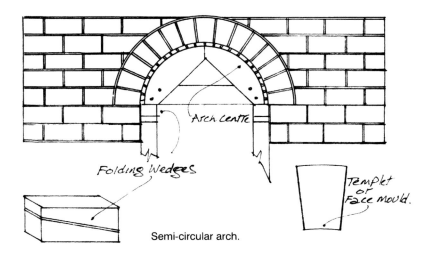

Semi-circular arch.

extrados rather than the intrados curve is that the greatest width of stone is here and this will, therefore, determine availability, ordering and so on.

3. Draw a perpendicular from the striking-point to the centre top of the arch. Set out the keystone, centred on this line.

4. Step out and mark the seven voussoirs on the extrados curve each side of the keystone.

5. Mark out the joint width, ie, 6mm; there should be sixteen joints.

6. Draw the fifteen voussoirs and their sixteen joints from the extrados towards the striking point, finishing on the intrados.

7. Make a templet in stiff plastic the same size and shape as one of the voussoirs (minus the joint). The same templet is used to mark all the voussoirs in the arch.

8. This templet is now used to mark the face shape of the voussoirs which are then pitched and their beds, and tops and bottoms squared off the face, similar to the cutting of a quoin stone.

9. As each voussoir is cut it is stood on the full-scale set-out on the floor and checked.

10. The arch centre is constructed of wood, well-braced and finished with narrow-width wooden laths laid and fixed to the curve with gaps in between. The position of each of the fifteen voussoirs is marked out on the centre (intrados curve), working from the full-scale set-out and using a dividers to ensure the keystone is exactly central.

    The centre is fixed in position on top of wooden supports with folding wedges underneath to make removal of the centre easy.

    The centre is levelled along its base and plumbed on its face. It is critical that the centre does not move as the arch is being built. Unless fixed securely centres have a tendency to go off plumb, bringing the voussoirs with them.

11. The building of the arch is commenced at both springing-points, laying each voussoir to the marked wooden centre.

    A string line or trammel is radiated from the striking-point of the arch to check that each voussoir is radiated correctly towards the striking-point.

A second line is used along the face of the wall to check that the arch is being built plumb, in line with the wall.

12. The last stone laid is the keystone (in vaulting, the 'boss'). It is important when laying voussoirs that mortar beds are spread in advance of laying, and not inserted afterwards when they will most likely fall out at a later date. In the case of the keystone the two adjoining voussoirs should be buttered with mortar, then the voussoir mortared on both its beds, so that when inserted it will squeeze into position tightly, compressing the mortar.

13. The wall either side and over the crown of the arch is built at the same time as the arch or at a later stage, depending on the size, stability and safety of the arch itself.

The cuts along the extrados line must be accurate and run with the curve. When cutting machinery is used for this purpose it is often noticeable how straight the cuts are, acting as tangents and not running with the curve of the extrados.

Assuming a joint size of 6mm there is little tolerance for error as the eye will notice the smallest deviation.

14. The timber arch centre is dropped by removal of the folding wedges. This should be done when the mortar has set quite hard, but at times is done fairly quickly on smaller arches so as to clean and point the underside of the arch. The centre to a semi-circular or Gothic arch can usually be dropped earlier than that of most segmental or semi-elliptical arches.

15. All voussoirs in the example given were the same size and shape, but often the keystone is larger and may project on the face, top and sides and be carved or lettered. This presents no problem except that the timber arch centre must be constructed with an allowance on top to take the dropped keystone.

Segmental arches

A segmental arch is less than a semi-circle or, in other words, its rise is always less than half its span. A common rise is $\frac{1}{6}$ the span, so a span of 4 feet (1200mm) has a rise of 8 inches (200mm).

Segmental arches are very common as face or relieving arches. They are probably the most commonly-used relieving arch over window and door openings, but they are also found in a whole variety of other places as well, including over heavy wooden beams on shop fronts hidden under renders.

The description given previously of the setting out, construction of the arch centre, stonecutting and laying are all applicable to the segmental arch. The only difference is the setting out, cutting and building of the skewback before the first voussoir is laid. This is done full-scale as before, with the arch set out and pre-cut in advance. There should be little difficulty in doing this as the procedure is self-explanatory.

What we will consider here instead is the more commonly found internal, rough, relieving arch in brick over a window, which will be later hidden under plaster.

1. The width of the opening is 4 feet 6 inches (1350mm) and, as in traditional work, a timber lintel or lintels have been placed across the opening on which to build the arch centre in stone/brick. Often these timber lintels are very slight in depth, sometimes only 2inches (50mm), as they have to carry only the small amount of masonry between their top and the underside of the arch. These timbers are sometimes found in a poor state and require replacement. New timbers should be treated before building in.

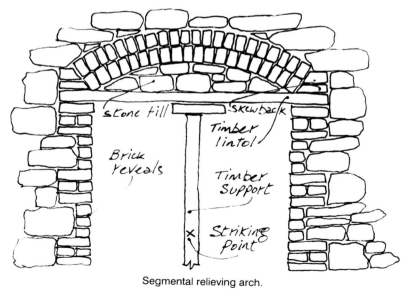

Segmental relieving arch.

2. The timber lintel/s are bedded into position with lime mortar and propped in the centre to prevent deflection, until the work overhead has hardened or set.

3. The striking-point of the arch will be somewhere down the vertical centre of the opening and should give a rise of about one-sixth of the span, which in this case is 9 inches or 225mm.

   The striking-point for rough, relieving arches is normally found quickly by trial and error. This can be done by trying various points down the vertical prop with a string line, until an arc can be drawn from the two springing-points reaching a point 9 inches above the springing-line. A nail is driven in at this point and the string line attached.

   In better-class work the finding of the exact centre is crucial. Given a span and a rise this can be done by bisecting the line from the springing-point to the top of the rise, and extending the bisector down until it meets the vertical centre line. This will be the striking-point of the arch.

   This is best done full-scale on a floor or sheet of wood.

4. Build the skewback, in this case using brick. This is done by first building then marking out the skewback on the brick, dismantling, cutting and re-building.

5. The centre to support the arch is constructed of brick or small stone, again using the string line to strike the curve.

6. The voussoirs, in this case bricks, are laid across the solid centre starting at each skewback, checking with the string line to their centres that they radiate to the striking-point. The bricks are not cut to shape but, instead, 'V' shaped mortar joints are used. Brick on edge is used in two rings rather than full bricks on end because the overall 'V' size of the joint is reduced.

7. The arch, as it is being built, is lined-in with the rest of the wall, using a second string line.

8. The work over the arch is now completed.

9. When the lime mortar has hardened, the props under the timber lintels can be removed.

Gothic arch

As mentioned elsewhere in this text the medieval Gothic arch in Ireland was often constructed of two stones only, and cut to shape with chamfers on the external arris.

The Gothic arch very efficiently transmits overhead weight vertically downwards without spreading, and thus requires less abutment either side of its span. This allows a reduction in the overall weight and thickness of masonry walls, particularly when it is used for vaulting.

The Gothic arch is set out from two centres (or striking points), has no keystone and comes in various styles:

- Equilateral (first appeared in Ireland during medieval times)
- Lancet (first appeared in Ireland during medieval times)
- Dropped (first appeared in Ireland during medieval times)
- Florentine (Gothic Revival from the nineteenth century onwards, Ireland)
- Venetian (Gothic Revival from the nineteenth century onwards, Ireland)
- Segmental (Gothic Revival from the nineteenth century onwards, Ireland)

Equilateral Gothic arch

The horizontal span and the radius to the intrados are the same.

The radii for both the intrados and extrados curves are struck from the two springing-points of the arch.

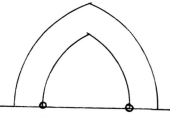

Equilateral Gothic.

Lancet Gothic arch

The radius to the intrados is greater than the span and, therefore, the striking-points always lie outside the span of the arch but always along the extended springing-line.

To find the striking-points given a span of 3 feet (900mm) and a rise of 3 feet 3 inches:

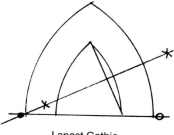

Lancet Gothic.

1. Draw the springing-line marking out the span of 3 feet and extending this line for a distance either side.
2. Draw a perpendicular from mid-span rising 3 feet 3 inches and beyond, and extend this below the springing-line as well.
3. Join the top of the rise with one of the springing-points and bisect this line, extending the bisector until it meets the springing-line.
4. We have now found one of the striking-points. To find the other, simply measure the same distance from the centre of the arch to the first striking-point, and do the same on the far side of the centre.

Dropped Gothic arch

The radius to the intrados is less than the span and therefore the striking-points are within the springing-points of the arch along the springing-line.

To find the striking-points follow the same procedure as before until the bisection of the chord (that connects the rise with one of the springing-points) reaches the springing line. (Step 3 above).

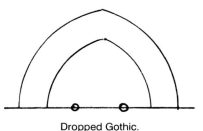

Dropped Gothic.

Camber arch

A very popular arch in the eighteenth and nineteenth centuries (much earlier in Italy). It derives its name from the small curved rise in the centre of $1/8$in per each foot of span (approximately 3mm per 300mm = 1%). This very small rise overcomes the optical illusion that a straight level line is dipped in the centre. The main advantage of the camber arch is that windows and doorframes did not have to be specially shaped on their heads to accommodate this arch. In some ways it looks and acts like a cork in a bottle. Vertical compression is converted to outward horizontal compression and therefore it jams in place.

Given the span, the skewback, the overall height of the arch and number of stones in the arch we can proceed.

1. Set the arch out to full-scale on a floor or sheet of plywood including the skewbacks.
2. The number of voussoirs is usually uneven to allow for a centre keystone.
3. The rise on the springing-line is drawn as a slight curve with a rise of ⅛in per 12in of span.
4. The uneven number of voussoirs are marked out on the extrados line from skewback to skewback, ensuring that the keystone is in the centre. The joints are also marked on the extrados line.
5. Along the intrados line mark out the same number of stones with joints. These will naturally be found to be narrower because of the shorter distance.
6. Extract templets from the full-scale set-out. Complete and cut voussoirs and skewbacks. A bevel is useful for measuring the odd angles which occur between the vertical beds and tops and ends of each voussoir.

Camber arch in brick. ⅛in rise per foot of span.

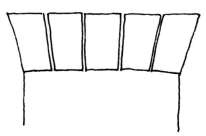

Camber arch in stone.

Setting out various other arch types

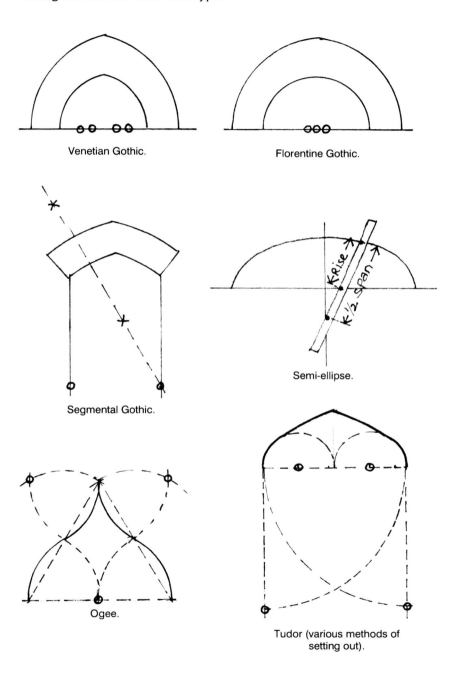

Venetian Gothic.

Florentine Gothic.

Segmental Gothic.

Semi-ellipse.

Ogee.

Tudor (various methods of setting out).

CHAPTER 8

QUARRYING

Where to get the stone we need is the next issue. Quarries for building purposes can be classified as rubble or dimension stone quarries.

RUBBLE QUARRIES

Rubble can be just about any stone that is usable for building walls. It can be igneous, sedimentary or metamorphic. There are very few stones that cannot be used for rubble walling, except those which are by nature very contorted or misshapen and do not lend themselves easily to being bedded or being roughly shaped.

Stone in many places lies just under the topsoil, and rubble stone can be extracted at times by simply digging a hole in the ground. This stone may be too fragmented or broken down from weathering to be of use, and it is only by going deeper that better stone is found. At other times rounded stones from ice flows are found near the surface and, although a little difficult, can still be used.

So rubble stone can be round, square, triangular, long and thin in shape, etc. In some sedimentary and metamorphic stones the faces are at an angle to the bedding planes and refuse to be worked or pitched by hand until square. These are useful 'as found' for building battered walls or laid with their slopes outwards like slates before being rendered.

Much of the rubble stone seen in the countryside for field boundaries and farm buildings was extracted during clearance while working the land.

Quarries, once opened to supply rubble stone for a particular project such as a church, or to build houses for a whole district, have mostly disappeared in recent years having been used for refuse tips or become filled with water.

Working in a limestone rubble quarry, County Limerick.

Quarries which extract rubble today supply the needs of the concrete and road-building sector, producing high-grade, washed and graded aggregates. The stone is extracted by blasting using explosives, and the emphasis is no longer on producing stone for building purposes. However, it is now common for 'stone pickers', as they have become known, to extract the best stone they can find in such quarries for building. Micro cracks caused by blasting can be a problem, leading to stones breaking unexpectedly when being shaped, or rainwater entering through capillary action on their faces.

Rubble stone in the past was a local indigenous material; it rarely travelled outside its own area.

DIMENSION STONE QUARRIES

The term dimension stone refers to the quality rather than the type of stone, so it can be any of the three classifications, i.e. igneous, sedimentary or metamorphic. A dimension stone is normally available in large sizes, is free from flaws, either structurally or aesthetically, and when it is being cut and worked it behaves in a consistent manner. The quarrying of dimension stone is a distinct industry and well-represented within the overall construction industry. Every year Carrara in northern Italy, famous for its white marble which Michelangelo worked, holds an exhibition on a world scale which features the latest technology and machines for the industry. In recent times, the industry has been revolutionised by computers, diamond-tipped saws and wire machines. Technology has reduced the cost of stone so that it can compete with other materials and because of this, and the fact that it is such a prestigious material for cladding and paving, especially for corporate and government work, it has returned to popular use in recent times.

In the 1960s and '70s stoneworking very nearly became extinct.

Dimension stone is also an international traveller – Irish limestone is very popular in Belgium where it is used for paving, tiling, cladding and so on. Similar to Belgian limestone which is commonly called black granite, Irish limestone is also exported to Germany and elsewhere. Dimension stone also travels from India and China around the world. Yorkshire stone from England has been sent to Ireland, worked to shape, and sent back to Yorkshire. Granite, quarried in Finland, has been processed in Spain and Ireland to be eventually fixed in position in London.

When a match could not be found locally for a limestone needed in Armagh city in Northern Ireland, a limestone from the USA was imported because it was the nearest equivalent.

Stone has been travelling for a long time – from 50 miles away to face the 5,000-year-old passage tomb in County Meath in Ireland, and from Bristol, England and Normandy to Ireland during the twelfth century to the fourteenth century for cut and carved work such as at Christchurch Cathedral, Dublin, and St Canice's Cathedral, Kilkenny. Water has always been the principal method of transport: sea, rivers and canals have all been used to transport stone and bricks.

So dimension stone is valued highly and is quite different from rubble stone. Let us now go to a typical, modern, dimension stone quarry, in this case limestone, and see how stone is extracted and processed there.

The scrub and topsoil are removed, together with varying amounts of inferior stone affected by weathering over the years, until a high-grade stone is reached. Limestone is laid down in beds – these were originally laid down as sediment together with the cementing agent, calcite, derived from the decomposition of sea organisms with bone and shell-type structures (see Chapter 5, Geology). Some of these beds are very deep. Vertical or near vertical cracks or joints occur every so often – a feature of limestone and resulting from shrinkage. Nature in its own way helps us remove a block of limestone by providing both horizontal and vertical joints, but to remove such a block it will be necessary to free the back end which is still attached. This can be done using plug and feathers or a channelling-machine which looks like a giant chain-saw.

Plugs and feathers have been used since Roman times, and probably earlier, to split and free stone. Nowadays, a series of holes is drilled using compressed air and a drill bit size up to 40mm diameter. The holes are spaced at regular intervals about 150/200mm apart and reach down to the next bed underneath. Into each one of these holes is placed two metal feathers with a plug in the middle. The plugs are driven in using compressed air, one set after another, until, eventually, the stone splits. Traditionally, the holes were drilled using a jumper chisel (bull-nosed at the cutting end) which was continually turned by hand and struck with a heavy hammer.

Plugs and feathers.

The plugs were also traditionally driven home with a heavy hammer. Before plugs and feathers, wedges of steel or iron were used. Wooden wedges were also used; these were wetted and allowed swell in order to lift or push a block of stone off its bed.

All sides of the block of stone are now free and a large loading-shovel can be used to dislodge the stone from its bed. The block of stone is then lifted by a crane to the cutting shed or loaded on to a truck.

Instead of plugs and feathers to free the back of the stone a large chain-saw (channelling-machine) is sometimes employed. This cuts a channel at the back of the block into which is placed a steel bag. Water is pumped into the steel bag which expands and moves the block of stone sufficiently off its bed to free it for lifting as before.

One of the most revolutionary machines in recent years to be used in the quarry is the diamond, segmented wire-cutter. Very similar machines were used earlier this century for cutting marble in other parts of the world, but grit was used then rather than industrial diamond for cutting.

A vertical hole – often 10 metres or more – is drilled down from the top bank or bed through numerous other beds, until the base of the quarry or a working level or table is reached. A horizontal hole is now drilled at the bottom in through the vertical quarry face until the vertical hole is reached. This is a skill in itself as the hole is easy to miss. The diamond-wire is now threaded through and joined as a continuous loop to a wheel drive mechanism at the base of the quarry. This motorised wheel is mounted on tracks travelling away from the vertical quarry face. On spinning, the wheel, moving away from the quarry face, pulls the revolving diamond-wire in a cutting action through the quarry face, much like a cheese-wire cutting cheese. Water is used to flush and cool the cutting-wire.

In Ireland, initially, these machines were used only for limestone quarrying but now have also been adapted to granite, although the wires have to be replaced more often on this harder stone.

When the wire breaks, a short new length can be inserted rather than replacing the whole wire.

After one cut like this is made the machine is moved on and a second cut made parallel to the first, so that blocks of stone may then be removed as explained above.

The face produced on the stone by the wire, although quite flat, is still not accurate enough for a dimension stone so the block of stone must now go through a secondary process in the cutting sheds.

The block is transported to the shed by an overhead gantry crane and mounted. There it is sliced by frame-saws, some of which have mono-blades in various widths to suit requirements. Large, diamond-tipped, circular saw-blades with a diameter of about 2.4 metres are also used. Computer technology allows these saws to run constantly day and night. Water is used to cool the blades.

Mono-blades can slice a large block of stone into multiple widths to look like sliced bread. These slices or slabs can vary from 20mm in thickness upwards (common thicknesses are 50mm, 75mm, 100mm, 150mm, 200mm, 225mm).

The next process is to take these slabs of stone and to square them up on a cutting bench using small-diameter, diamond-tipped saws. Machines like these will produce six-sided stone with each face 'out of twist' and square to the next face.

Such accurately cut stone can now be further cut, chamfered, or moulded.

The surface dressing of stone is also automated to produce various finishes.

The cutting process with stones such as granite, marble, sandstone and so on is similar.

CHAPTER 9

STONEMASONRY

ANATOMY OF A STONE WALL

In order to understand stone buildings we must closely examine the anatomy of a stone wall. Stonemasons everywhere agree that for stone walls which are structurally stable you need the following elements:

Foundations (solid loadbearing ground and a foundation or wall of sufficient thickness to spread the load of the building over this ground without sinking or differential settlement).

Double wall (two wall faces), one on the exterior and the other on the interior of the building. These wall faces can be laid as rubble or of cut stone. Lime mortar is normally essential to bed these faces, but mud is also used. (Dry-stone structures exist and are structurally sound, but to build to any height with an economic wall-thickness that will keep out the weather lime mortar is essential).

Through and bond-stones to tie the wall transversely (across its width).

Core or hearting of stone and lime mortar (sometimes mud) in the centre of the wall.

Quoin stones at corners and openings, cut accurately to shape or just roughly shaped, depending on the quality of work.

Foundations

Foundations, where they existed, were based on empirical knowledge of what had worked in the same locality with similar wall-thickness, height, etc. Where this was not known the nature of the ground being excavated was studied and, based on experience of similar soils, a judgement was made about what to do. In many cases, the wall-thickness of the building was such that it could easily spread the dead and live loads of the building over the

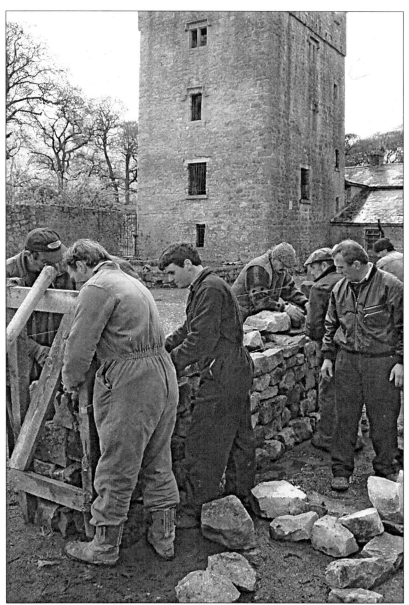

A stonewalling workshop in County Kildare.

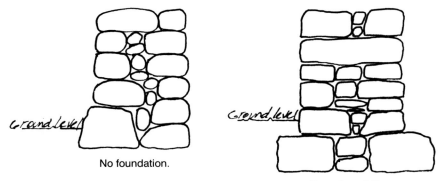

No foundation.

Stone foundation.

ground, without sinking or differential settlement. At other times, foundations projected each side of the face of the wall. In some cases where there was soft ground, piling and planking with wood was carried out in advance of the wall being built.

In the nineteenth century and possibly well before, inverted arches were used. These can be seen at or below ground-level overcoming differential settlement by spreading isolated loads over a greater ground area.

Traditionally, heavy stone buildings were given a generation to settle.

The best foundation was, of course, solid rock.

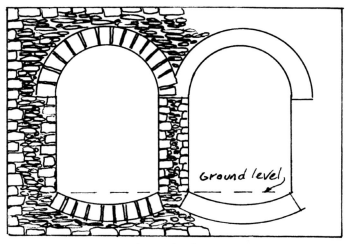

Inverted arches.

Traditional face bonds

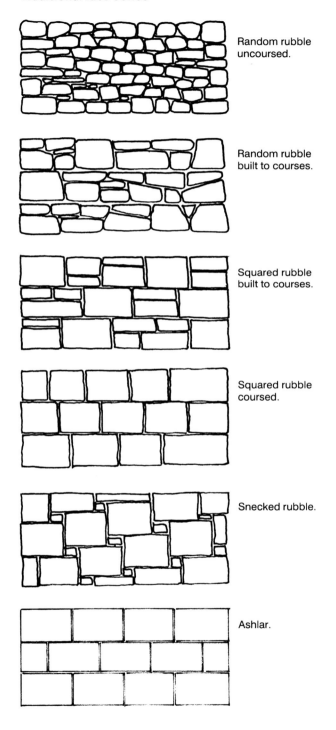

Random rubble
uncoursed.

Random rubble
built to courses.

Squared rubble
built to courses.

Squared rubble
coursed.

Snecked rubble.

Ashlar.

Double walls

Solid double walls are standard practice on most stone buildings – an outer and inner stone face, usually separate, and occasionally tied together with through-stones (see below). The stone on these faces may be simply rubble stone laid as found with fairly large joints (reduced with pinnings, discussed later). Alternatively, variations of cut stone, some of which may have flat-tooled, or decorative faces with small joints were sometimes used. Double walls may have a cut stone external face and a rubble internal face and may be exposed on both faces, or, alternatively, rendered externally and plastered internally with lime mortars. The width of a wall in relation to its height is crucial – if this ratio becomes too great then a wall is in danger of collapse or of bending under applied loads. Rarely is this an issue with stone walls as, in general, they are of ample width in relation to their height. Floors and roofs with cross ties/rafters connected to other rafters all restrain walls from collapsing. Double walls are often reduced in thickness on their inner face as they ascend with these offsets used to support floors. At other times, offsets are to be seen externally. They are used in a decorative manner, as a string course of cut stone which is chamfered. Vaulted floors, of course, require walls of sufficient thickness and mass to resist their outward thrust.

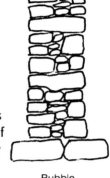

Rubble.

Thick rubble wall.

Through- and bond-stones

In order to tie a double wall across its width (transversely) through-stones were used. A through-stone reaches across the width of the wall from one face to the other and, therefore, is usually to be seen on walls with a thickness of c. 600mm and less, rubble stones exceeding 600mm in length being difficult to find in normal circumstances. Good practice was to use one through-stone

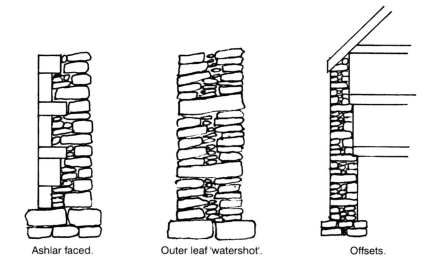

Ashlar faced.　　　　Outer leaf 'watershot'.　　　　Offsets.

per m² of vertical wall face and to stagger one over the other. Where through-stones are not to be found bond-stones, which travel a considerable distance ($c.^2/_3$) across the width of the wall, are used. These are not only staggered vertically as before but also horizontally as well, so that a bond-stone laid from the external face of the wall in, is followed by a bond-stone laid from the internal face of the wall out. On very thick walls all face stones, as far as possible, should be laid with their length in to the wall in order to bond these faces into the core. These very thick walls usually lack any real transverse bonding and therefore are subject to loss of wall faces and bulging. In many cases, nineteenth-century interventions can be seen which were produced by the local blacksmith – round iron bars as tie irons with 's' shaped anchors at ends to hold the walls of stone buildings together.

Cores or hearting

The centre of stone walls generally consists of stone and lime mortar (or mud). In the worst of cases this appears to have been shovelled in without any care, with stones lying vertical acting as wedges and excess amounts of lime mortar or mud. This original carelessness is often the cause of problems that now require structural intervention. Cores should be bonded properly with stones laid on their natural beds (if sedimentary) and with minimum joint sizes. Hot lime mortars are evident in many cores, presumably for fast drying of lime mortars used in bulk, which otherwise would slow down production.

Quoin stones

The corners of a building are like the corners of a cardboard box – once lost they leave the sides vulnerable to collapse. Quoin stones occur at external corners of buildings and on the jambs of various openings. They may be cut accurately to square and be face dressed or simply laid as found, being reasonably square in shape. Often in rubble work they are simply shaped up with a hammer and pitcher. Well-cut quoin stones are sometimes a little less than square to avoid snagging the line when it is stretched along their length. On plan they are sometimes triangular in shape to save on transport costs and to reduce their weight for lifting into position, although this should not be considered good practice.

Quoin stones are not always square in shape and are sometimes cut at various angles to suit a particular geometric shape on plan. They may also be rebated, chamfered and decoratively worked. The length of a quoin stone is important in order to tie corners together; structurally, the longer the better. Similarly, internal corners should be well-bonded with long stones.

Face bond

Vertical joints, if excessively long on wall faces, cause failure and collapse. Excessive vertical joints are the result of putting a series of stones on top of one another in such a way that their vertical end joints (called perps) align. The mythical Irish stonemason, the Gobán Saor's rule was 'I on 2 and 2 on I' which is the best practice though not always achievable without undue cutting. For rubble work '2 but not 3' is more comfortable, but if two rather high stones are placed one directly over the other then an excessively long vertical joint will result which is not acceptable.

Facework is sometimes 'watershot', ie, the beds tilted out to the external face. This is possible with 'slab-like' stones such as slate but can be very difficult and even impossible with other stones. The tilt of watershot stones can produce a ledge on the external face of the wall on which to hang renders. This is a good reason for leaving renders in place and not removing them to expose the stone, as the ledges can trap water.

1 on 2 and 2 on 1.

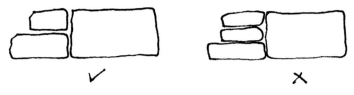

2 but not 3.

Bed bonding

Good bonding on wall beds is achieved by running the lengths of face stones into the heart of the wall. If possible, use through-stones generously at 1 per m² of facework and vertically staggered. Alternatively, 2 bond-stones per m² of facework, again staggered vertically, but also horizontally from one face to the other. Tilt through-stones and bond-stones slightly out to the external face, keeping overall volume of mortar down to improve stability and to reduce expenditure.

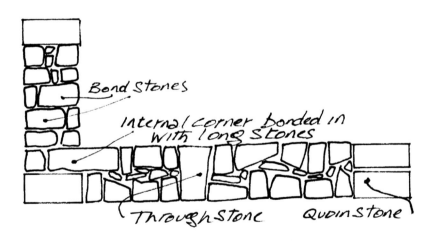

Bed bonding.

Mortar joints and pinnings

In good work joint sizes are consistent and kept to a minimum. In cut stone this is relatively easy but in rubble work joint sizes may vary considerably because of uncut stone shapes. Of course rubble work is often roughly shaped and as a consequence joint sizes are reduced. As discussed elsewhere, (Chapter 12, Lime), lime mortar carbonates most effectively when used in small thicknesses. So the plasterer uses coats not exceeding 12mm while the stonemason endeavours to reduce joint sizes in rubble work. Throughout the world this is achieved with stone pinnings – pieces of stone inserted into fresh mortar joints using a hammer to secure them at the end of each day's work. Large vertical and horizontal joints are dealt with this way. In the case of horizontal joints the pinnings are tilted slightly outwards. The end result is an overall general thickness of mortar approximating 12-20mm.

These mortar joints also display other small stones, often flat in shape, which are used to prevent a large overhead stone rocking about or to reach course heights. These stones are laid during the work and not introduced later so they are locked in position securely by the compression of overhead stones. The loss of the stone pinnings inserted into fresh mortar at the end of each day is always a strong indicator that pointing is necessary.

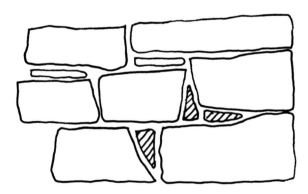

Stone pinnings inserted at the end of a day's work illustrated by shading.

Natural bedding

Sedimentary stones (limestone and sandstone but also metamorphic stones which display bedding planes like mica schist, quartzite, etc.) should be laid on their natural beds and never face-bedded. Occasionally edge-bedding is necessary for copings on walls and also for sills, barge stones and so on, particularly with stones like sandstones which will de-laminate with weathering. Not all stones display natural bedding planes which are visible.

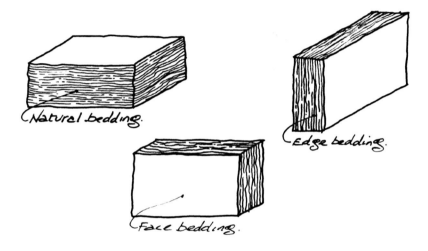

Balance

Stones should be laid level, with their bed length on the face of the wall greater than their heights, even if uneven in shape.

Bed length should always be greater than the height of the stone.

Stones should be laid in balance and not at unnatural angles.

Roughly built vernacular dwelling.

Lime mortars

Lime mortars are discussed in detail in Chapter 12 but for stone-masonry they were usually of the hot lime variety with coarse sand (rubble work).

Sequence of construction

Good construction requires the walls of a building to rise evenly all around and not for one side of a building to far exceed the other side in height. This is to prevent unequal loading of the ground surface which can lead to cracking in masonry structures. Internal cross walls should be bonded into external walls (as the work proceeds, if possible).

CHAPTER 10

QUANTITIES AND ESTIMATING

And now for the cost! In recent years building with stone has come back from near extinction. People ask how much material is required to carry out a job, how long it will take and what will it cost? Questions of time and cost are far from easy to answer for the following reasons:

- A general lack of skilled and experienced stonemasons and contractors in stone.

- Very little understanding of stone amongst the professions.

- A general lack of good-quality rubble-stone quarries throughout the country, with most being simply blast quarries for the production of aggregates. Rubble stone can also vary greatly within a quarry – some can be quickly loaded onto trucks while more needs to be picked by hand.
  At the other end of the scale the dimension-stone industry is well organised and produces high-quality cut stone.

- Very broad descriptions of some types of work, such as rubble stone, coursed or uncoursed, can mean a variety of things to a variety of people. The maximum joint size, whether pinnings can be used in joints or whether joint sizes must be uniform and small, how easily the stone can be cut and shaped or whether the stone can be laid nearly as found – all have an enormous effect on time and cost. If work is not specified accurately and if the standard of work required is not clearly understood then disputes will ensue. Yet a high standard of finish may be impossible to achieve because of the nature of the stone to be used.

- Very little time is allowed prior to the commencement of many contracts for stone to be quarried and cut. There is a general belief that all quantities, sizes and shapes are readily available.

- The insistence that stone be a uniform colour, have no fossils, and at times be a particular size, bordering on the maximum possible within the beds of the quarry, increases costs and makes stone costlier than other materials. Stone is a natural material and to reveal its beauty one must work with it and not against it.

QUANTITIES

The following is only an introduction to the whole field of quantities and much information has been excluded in order to simplify the process.

Quantity surveying is a profession concerned with the cost of building. An architect designs a building, producing drawings and specifications. The drawings show the plan, elevation and sections while the specification describes the type and quality of material and work required to complete the building.

The quantity surveyor working from this information produces a bill of quantities. The bill of quantities is, within reason, a precise measurement and description of the materials and work necessary to construct the building.

Work is described and measured in a particular way by the quantity surveyor so that the builder will understand clearly what is intended. This system of measurement is called the Standard Method of Measurement or SMM for short.

A number of builders are furnished with the drawings, specification and bill of quantities, and asked to price the work. They then submit a tender by pricing each measured item in the bill of quantities, resulting in an overall tender figure.

Computer programmes to assist both the builder and the quantity surveyor in the preparation of quantities and estimates have revolutionised the process. The examples given here are all calculated by hand – a method suitable for small projects only.

Stonework in general is measured on the face by the square metre (m^2) stating the thickness. So, for instance, a wall 8 metres long, 3 metres high and 600mm thick would be described as $24m^2$ by 600mm thick.

The quantity surveyor always writes measurements in the order: length, width, and height, or LxWxH. Width can also be referred to as breadth or thickness while height can be referred to also as depth. This excellent method of work – avoiding confusion and allowing work to be more easily checked at any time in the future – is also useful in other areas such as the ordering of dimension stone.

Specially ruled paper called dimension paper used for taking off quantities has four vertical columns called times, dimension, square and description.

So if we had three walls as previously described, 8 metres long, 3 metres high and 600mm thick built as random rubble (uncoursed) in 1:2:9 cement: lime: sand mortar this would be taken off as follows:

T	D	S	Description
3/	8.00		
	3.00	72 m^2	Random rubble wall (uncoursed) 600mm thick built in 1:2:9 mortar.

The quantity surveyor converts this to a different format in the bill of quantities which the builder receives, by only showing the final measurement as seen in the square column above, in this case 72m^2 plus the description. If there is other similar work in the contract then this may be accumulated so as to show a grand total – a process called abstracting. Any work that differs from the normal such as attached piers, plumbing and cutting required at corners, window and door opes will all be measured separately in the bill of quantities in lineal metres to allow the builder to consider the additional work required and price accordingly.

The above example would appear in the bill of quantities like this:

Stone masonry – random rubble wall (uncoursed) 600mm thick in 1:2:9 mortar 72 m^2.

Let us now move a stage further, taking three stone walls 8 metres long by 3 metres high and 595mm thick. This time we take into account that there are concrete foundations 1200mm wide by 500mm deep and that the distance from ground level to the bottom of the concrete foundation is 800mm. From the top of the concrete foundation to ground level the stone wall is built in 1:1:6 cement, lime and sand, while above ground level it is built in 1:2:9.

The wall is finished on top with a cut half-round coping 595mm wide built in 1:2:9 mortar. In addition, all four corners of each of the three walls have quoin stones measuring 595mm x 295mm x 295mm bedded with 5mm joints and the random rubble is now built to 300mm courses to match the height of the quoin stones.

In taking off quantities, topsoil is normally accepted as 150mm deep and measured separately from the excavation of the foundation trench because it will be reinstated at a later date – this has been deleted here for the sake of clarity. Trenches are measured in stages according to depth so that the builder can allow for additional work (including planking and strutting which may be necessary to the trench sides) and safety measures required as the trench gets deeper (for instance, 'not exceeding [n.e.] 1.5 metres in depth', 'exceeding 1.5 metres but not exceeding 3 metres', etc.)

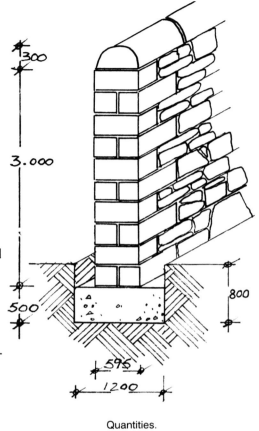

Quantities.

Similarly concrete is described in the bill of quanties as n.e. (not exceeding) 150mm thick, over 150mm but n.e. 300mm thick, and over 300mm thick to allow for differences in pricing.

T	D	S	DESCRIPTION
3/	8.00 1.20 0.800	23.04 m³	Excavate foundation trench n.e. 1.5 metres deep.
3/	8.00 1.20 0.50	14.40 m³	concrete to foundations over 300mm thick.
3/	8.00 0.30	7.20 m²	random rubble wall, below ground level (built to courses) in limestone 595mm thick in 1:1:6 cement lime and sand.
3/	8.00 3.00	72m ²	rubble wall (brought to 300mm courses) in limestone 595mm thick laid in 1:2:9 mortar with joints finished flush both faces.
3/ 2/2	3.00	36 lin m	extra over random rubble (brought to courses) for quoin stones in dimension limestone with punched finish in sizes 595mm long x 295mm wide x 295mm high laid in 1:2:9 mortar with 5mm joints.
3/	8.00	24 lin m	half-round dimension limestone coping 595mm wide (297.5mm radius) with punched finish laid in 1:2:9 mortar with flush joints.

As shown above the dimension quoin stones are measured as extra over and above the cost of doing it in rubble (3/2/2/ 3.00). This simply means three walls x two ends, each wall x two quoins, each end x the height of the wall above ground level.

One of the items omitted for the sake of clarity is the backfilling of the trenches on completion of the work and the reinstatement of the topsoil – the amount of soil remaining to be carted away is not included.

The examples given were for straightforward walls. In taking off quantities on buildings corners must be reckoned – here centre-line measurements give the true length of trenches, foundations, walls, etc.

Take a simple rectangular building 12 metres long and 9 metres wide externally, with walls 400mm thick. The centre line or true length of this wall is found by first calculating the external perimeter distance.

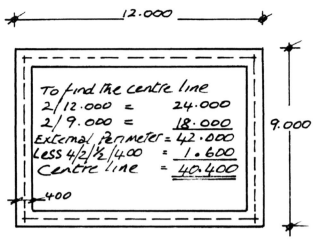

To find the centre line.

2/12.00	24.00
2/12.00	18.00
Ext. Perim. =	42.00 metres

The centre line is now found by deducting 4/2/½/400 = 1.60 metres.

42.00 - 1.60 = 40.40 metres = centre line of wall (also of trench and foundation.)

Just measuring the face of the wall (42 metres) does not give an accurate length and results in an error of 1.6 metres which means the tender will be overpriced and therefore lost to somebody measuring more accurately.

Using 4/2/½/400 is the simplest method of finding the centre line. This represents measuring half the thickness of the wall (twice) at each of the four corners. This is best understood by looking at the drawing above.

All buildings, no matter how many right-angled corners they have, will always have four more external corners than internal corners on the external face of the wall. Once again, best understood with the aid of a drawing (page 124). This means that the external length of the perimeter can be calculated by the addition of external measurements and then a deduction of 4/2/½/thickness of the wall made to find the centre line of the wall. The alternative to this method – the

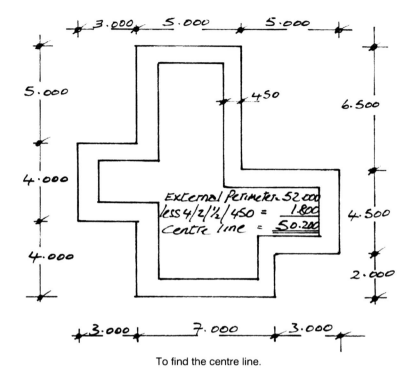

To find the centre line.

adding and subtracting of many measurements – would be difficult and increase the chance of error. In quantities, the fewer moves one makes to reach an answer the less chance of error.

The drawing above shows a building with eight external and four internal corners on the exterior face. The building has an external perimeter length of 52 metres and a wall thickness of 450mm.

To find the centre-line measurement

External perimeter length	=	52.00	
Less 4/2/½/450mm	=	1.80	
Centre line	=	50.20	

The internal wall perimeter of the same building can be easily found given only external measurements. This is useful for calculating areas of internal plasterwork, painting, etc.

External perimeter length   =   52.00
Less 4/2/450mm   =   3.60
Internal perimeter length   =   48.40

In this case we took the four corners but came in twice the width of the wall at each corner to reach the internal face of the wall.

Openings such as windows and doors are deleted later when the quantities for these elements are being taken off.

ESTIMATING

Estimating is a science based as far as possible on accurate verifiable information which, given the same circumstances and conditions, will repeat itself. An experienced estimator will take this known information and adjust it to suit new situations – the more experienced the estimator the more accurate the judgement.

Estimates that lean on the side of too much caution will be too high and result in no work; estimates that are overconfident or not based on true information may be too low and result in work being won but a loss being made.

An estimator should have available historical cost data in relation to material and plant costs but also, more importantly, just how long it takes to carry out a unit of work such as a m^2 or a m^3.

To turn an estimate into a tender the owner or management of a firm weighs up factors such as the risk involved, the volume of work currently or soon to be in hand, the availability of skilled personnel and so on. The estimate may be adjusted upwards if the work is not keenly desired or lowered if badly needed. The result is the tender or how much the contractor will charge to carry out the work.

The accurate preparation of estimates is time-consuming and costly to the contractor and therefore it is not good practice to invite more than three contractors to prepare a tender.

As seen earlier in the quantities section many elements of work are measured by the lineal, square and cubic metre (lineal M, M^2, M^3). The contractor builds up unit rates for these which usually include material, plant, labour, overheads and profit. Overheads include the cost of supervision, insurances, administration and many other elements. Profit is what the contractor expects for a return on capital investment and the risk taken. Often profit is quite small, and if a job is priced too low, the profit margin is the first to suffer as the overhead costs, etc. must still be met.

The following unit rates include material costs only. These will be for a number of elements related to working with stone and lime mortars. They should only be taken as examples and presumably will soon be out of date after publication. Adjustment should be made for the cost of material in your own area.

Mortar

Traditional lime mortars

Lime-only mortars are making a comeback in architectural conservation for repairs to stone walls, pointing, plastering and rendering. They are also being used in new work executed in a traditional manner.

Probably the most common mix used in this work is 1:3 or 1 lime to 3 sand.

This is because the void space in well-graded sand is 33% and the lime therefore replaces the air content – when 1 m³ of lime putty is mixed with 3 m³ sand we get 3 m³ of mortar instead of 4 m³. The term used to describe this is 'yield'.

Lime putty is available in 25kg. tubs and weighs approximately 1.25 tonnes per cubic metre.

Coarse sand weighs approximately 1.60 tonnes per cubic metre, dry. The currency used is euros (€).

For 1:3 lime putty and sand mix

The material cost of a 1 lime putty to 3 sand mix may be worked out as follows

1 m³ lime putty	=	1.25 tonnes
$\dfrac{1.25\ tonnes}{25\ kgs}$	=	50 tubs of lime putty

Cost per m³ of a 1:3 lime putty and sand mix (5mm max. particle size)

1 m³ lime putty = 50 tubs of lime putty @ €15 (incl. VAT)	=	750.00
3 m³ sand @ 1.6 tonnes per m³ = 4.8 tonnes		
@ €15 (incl. VAT)	=	72.00
Add 2.5% waste on lime putty	=	18.75
Add 10% stockpile waste on sand	=	7.20

| Material cost for 3 m³ of 1:3 lime putty mortar | = | 847.95 |
| " " " " 1 m³ " " " " " | = | **€282.65** |

Cost per m³ of a 1:1 lime putty and fine sand (1mm max. particle size)

A 1:1 mix is used for the finish coat on internal plasterwork.

Cost per m³ of a 1:1 lime putty and fine sand mix (max. sand particle size 1mm)

1 m³ lime putty a.b.	=	750.00
1 m³ fine sand (assume it weighs 5% less than coarse sand = 1600kgs less 5% = 1520 kgs per m³) @ 1.52 tonnes per m³ @ €30 per tonne (incl VAT)	=	45.60
Add 2.5% waste on lime putty	=	18.75
Add 10% stockpile waste on sand	=	4.56
Material cost for 1.66 m³ (sand has 0.33 yield)	=	818.91
" " " " 1 m³ of 1:1 lime putty and fine sand	=	**€493.32**

Hydraulic lime mortars

Cost per m³ of a 4:5 mix of 3.5 natural hydraulic lime and sand

3.5 NHL costs €19 (including 21% VAT) per 25 kg bag or €760 per tonne
3.5 NHL weighs 600 kgs per m³

4m³ of 3.5 NHL @ 600 kgs per m³ = 2.4 tonnes @ €760	=	1824.00
5m³ sand @ 1.6 tonnes per m³ = 8 tonnes @ €15 (incl. VAT)	=	120.00
Add 2.5% waste on 3.5 NHL	=	45.60
Add 10% stockpile waste on sand	=	12.00
Material cost for 7.33 m³ (allows 0.33 yield on sand)	=	2001.60
" " " " 1 m³ of a 4:5 mix of 3.5 NHL and sand	=	**€273.07**

Cost per m³ of a 1:2 mix of 3.5 natural hydraulic lime and sand

1m³ of 3.5 NHL @ 600 kgs per m³ = 0.6 tonnes @ €760	=	456.00
2m³ sand @ 1.6 tonnes per m³ = 3.2 tonnes @ €15 (incl. VAT)	=	48.00
Add 2.5% waste on 3.5 NHL	=	11.40
Add 10% stockpile waste on sand	=	4.80
Material cost for 2.33 m³ (allows 0.33 yield on sand)	=	520.20
" " " " 1 m³ of a 1:2 mix of 3.5 NHL and sand	=	**€223.26**

Cost per m³ of a 1:2.5 mix of 3.5 natural hydraulic lime and sand

1m³ of 3.5 NHL @ 600 kgs per m³ = 0.6 tonnes @ €760	=	456.00
2.5m³ sand @ 1.6 tonnes per m³ = 4 tonnes @ €15 (incl. VAT)	=	60.00
Add 2.5% waste on 3.5 NHL	=	11.40
Add 10% stockpile waste on sand	=	6.00
Material cost for 2.67 m³ (allows 0.33 yield on sand)	=	533.40
" " " " 1 m³ of a 1:2.5 mix of 3.5 NHL and sand	=	**€199.77**

Rubble walling

Rubble stone can be handpicked from the quarry and transported back to site causing little waste but considerable expense in time and labour.

Alternatively, rubble stone can be selected by loading-shovel and delivered to site with quite an amount of waste, a saving in labour by not going to the quarry – but extra time is required to rough-shape stones.

In either case it is best to find the right rubble-stone quarry to begin with. Most rubble-stone quarries – set up to produce aggregates of various grades and sizes for concrete work and road-building – are not interested in extracting large sizes with flat beds and reasonable faces, but in most, useful stone is found lying in large heaps after a blast has taken place.

Rubble stone delivered to site by waste disposal skip

12 tonnes of rubble ex-works (incl. VAT) @ €7 per tonne	=	84.00
Waste disposal skip hire	=	90.00
Total for 12 tonnes	=	174.00
Per tonne of rubble stone	=	**€14.50**

Remember to add labour for selecting at quarry.

Machine selection of rubble stone and delivery to site

A general price per tonne for selection by loading-shovel and delivery to a site within reasonable distance is €30.00 per tonne (incl. VAT).

The amount of waste varies according to the natural quality of the quarry itself and the skill of the machine-operator in selecting a good area within the quarry.

Assuming that care has been taken in selecting a quarry and the machine-operator understands exactly what is needed, you should still allow 25% to 33% waste.

Cost per tonne selected by machine and delivered to site (incl. VAT)	=	30.00
Allow 25% waste	=	7.50
Total per tonne for machine-selected rubble including delivery	=	€37.50

We will take the figure of €37 per tonne for estimating purposes. A rubble stone wall has from 25% to 33% mortar by volume. The area that this lime mortar takes up within the wall means that there is a saving in the same volume of stone, but for estimating purposes this is not considered and is accepted as waste. The calculations below are based on a mortar volume content of 25%.

Irish limestone and granite each weigh 2.70 tonnes per m³.

Cement: lime: sand mixes such as 1:1:6 and 1:2:9 are used to build modern walls and cost c.€100 per m³.

For 600mm thick random rubble wall (uncoursed) in 1:1:6 or 1:2:9

Stone 1m x 1m x 0.60m x 2.70 tonnes = 1.62 tonnes @ €37 per tonne	=	59.94
Mortar 25% of 1m x 1m x 0.6m = 0.15 m³ @ €100 m³	=	15.00
Material and plant cost per m² random rubble wall (uncoursed) in 1:1:6/1:2:9	=	€74.94

To build this same wall in lime putty and sand (1:3) will cost the following

1:3 lime putty and sand @ €282.65 per m³ a.b.

0.15 m³ of 1:3 mortar @ £282.65 per m³	=	42.40
Previous cost of 1:1:6/1:2:9 mortar	=	15.00
Difference	=	+ €27.40

Material cost of 600mm thick random rubble wall (uncoursed) in 1:3 lime putty and sand per m² = €74.94 + €27.40 = **€102.34**

For 450mm thick random rubble wall (uncoursed) in 1:1:6 or 1:2:9 mix

Stone 1m x 1m x 0.450m x 2.70 tonnes per m³ = 1.215 tonnes @ €37.00 per tonne	=	44.96
Mortar 25% of 1m x 1m x 0.45m = 0.113 m³ @ €100	=	11.30
Cost per m² for 450mm thick random rubble wall uncoursed in 1:1:6/1:2:9	=	**€56.26**

To build the same wall in lime putty and sand (1:3) will cost the following

1:3 lime putty and sand @ €282.65 per m³ a.b.

0.113 m³ of 1:3 mortar @ €282.65 per m³	=	31.94
Previous cost of 1:1:6/1:2:9 mortar	=	11.30
Difference	=	**+ €20.64**

Material cost of 450mm thick random rubble wall (uncoursed) in 1:3 lime putty and sand per m² = €56.26 + €20.64 = €76.90

Internal plasterwork

Hair

Traditional internal plasterwork uses animal hair in the first two coats (scratch and float), and specifies 9 lbs of hair per cubic yard = approximately 5.5 kgs per m³. This, in practice, seems excessive and one wonders if this specification was, in fact, ever followed. Ox hair was traditionally used, but nowadays both goat and yak hair are imported for this purpose.

Allow 3 kgs per m³ of mortar @ €13 per kg. (incl. VAT)

Traditional internal plasterwork in three coats of lime putty and sand with hair on a stone background

Nominal overall thickness of three coats will be c.23mm.

Earlier we established the material cost of 1 m³ of 1:3 lime putty and sand as €282.65 and 1:1 lime putty and sand as €493.32 per m³.

Pointing

Pointing, if necessary, should be estimated separately and is not allowed for here.

Daubing out

Daubing is the filling in of hollow spots on the face of a wall. On rubble stone the necessity to do this can vary widely. Where hollow spots are quite deep they should be filled with multiple coats of 1:3 lime putty and sand with hair, in coats not exceeding 12mm in thickness. In this case, clay brick or stone spalls should be used on the flat to reduce the overall thickness of the daubing out.

The quantity of material required to daub out hollow areas on stone walls prior to plastering will vary greatly according to the nature of the wall face behind.

Some allowance has been made for daubing out in the following analysis by allowing 18mm in thickness for the first or scratch coat which has a nominal thickness of only 12mm.

Scratch coat

Material cost of 1:3 lime putty and sand per 1 m³   =   €282.65

The scratch coat is estimated as 12mm thick but undulations on the face of the wall, compression and waste will result in a necessary allowance of 18mm.

1 m³ applied as 18mm thick will cover approximately 55 m².

Material cost per m² for scratch coat = €282.65 / 55	=	€5.14
Hair @ 3 kgs per m³ @ €13 per kilo = €39 / 55	=	€0.71
Material cost per m² for 1:3 lime putty and sand scratch coat	=	**€5.85**

Float coat

The float coat is estimated at 8mm thickness but allow 10mm for undulations, compression and waste.

1 m³ of 1:3 mix applied as 10mm thick will give 100 m²

Material cost per m² for scratch coat = €282.65 / 100	=	€2.83
Hair @ 3 kgs per m³ @ €13 per kilo = €39 / 100	=	€0.39
Material cost per m² for 1:3 lime putty and sand float coat	=	**€3.22**

Finish coat

The finish coat is estimated at 3mm thick but allow 5mm for slight undulations, compression and waste (no hair in finish coat).

1 m³ of 1:1 mix applied 5mm thick will give 200 m²

Material cost per m² for scratch coat = €493.32 / 200   =   **€2.46**

Material cost for 3 coat work in lime putty and sand with hair per m²

€5.85 + €3.22 + €2.46   =   **€11.53**

Internal plasterwork on laths in three coats a.b. with hair

Plastering on laths was common practice throughout much of the world until well into the twentieth century until sheet plasterboard and dry-lining boards took over. In old houses plastering of ceilings and partitions and the internal lining of both stone and brick walls was often carried out with timber studding, using laths plastered in lime putty and sand with animal hair.

The differences between plastering on a stone background and a wood-lath background are

- There is generally a quicker drying-time on the stone background.
- The first or scratch coat on a wood lath background requires additional material to form the key between the laths – can be excessive if the material applied is too wet or if too much trowelling is used.
- Two-coat work may be possible on machine-cut laths because of the reasonably flat, plane surface created.

Laths

Laths were traditionally split by hand along the grain of the wood although machine-cut laths were introduced sometime in the later nineteenth century. Laths were about 25mm wide, 10mm thick for ceilings and say 8mm for partitions fixed with iron nails.

A saw-mill will not differentiate much between these thicknesses and will charge about 5p per foot or 17p per lineal metre (incl. Vat). Laths are now best fixed with galvanized clout nails say 25mm long with approximately 6-8mm gaps between laths. Stud timbers or ceiling joists are traditionally fixed at 300mm centres.

To find the amount of lineal metres of lath to fix 1 m²

$$1\ m^2 \quad = \frac{1m \times 1m}{(25mm + 6mm)} \qquad = 33\ \text{lin. metres}$$

33 lin. metres per m² @ €0.08	=	€2.64
Waste @ 5%	=	€0.13
Galvanized clouts €6 per kilo, nails 25mm long @ *c.* 300 per kg.		
	=	€2.40
Waste on nails @ 5%	=	€0.12
Laths and nails per m²	=	**€5.39**

Laths should be pre-wetted before applying the scratch coat.

The previous analysis for three-coat work on stone can be used here.

The allowance of 18mm of material for the scratch coat to achieve a nominal thickness of 12mm should be adequate.

Material cost a.b. for three-coat work in lime putty and sand with hair per m²	=	€11.53
Material cost per m² for three-coat lime putty and sand including timber laths and nails = €11.53 + €5.39	=	**€16.92**

External rendering

The mixes for external rendering are very much dependent upon the nature of the sub-strata, the location of the site and the location of the element or area of a building which requires rendering. Therefore non-hydraulic mixes may be acceptable in one instance but a hydraulic mix may have to be used in another.

As discussed with internal plasterwork, pointing and daubing out may be necessary before commencing.

Traditional wet-dash or harling

Wet-dash, traditionally, was applied as one or more coats to buildings, boundary walls and so on. The material used – coarse sand and lime – was often identical to the mortar used to build the wall.

Traditional practice was to throw on all coats, unlike modern practice which is to scud, scratch, maybe float and then to wet-dash.

Good practice in external renders is to work out in ever weaker coats. Working with natural hydraulic lime with a grade of 3.5 NHL, we will apply 3 coats. The first two at 1:2 and the last at 1:2.5.

Material costs of a 1:2 mix of 3.5 NHL and sand per m³ as before

= €223.26

Material costs of a 1:2.5 mix of 3.5NHL and sand per m³ as before

= €199.77

Allow say 10mm per coat. Therefore three coats will be 30mm, but with compression etc. this will average something just over 20mm when finished.

The first two coats at 1:2 in 3.5NHL @ 10mm each = 20mm. 1m³ of material at this thickness will cover 50m². €223.26/50   =   €4.46

The last coat at 1:2.5 in 3.5NHL @ 10mm thick will cover 100m². €199.77 / 100   =   €2.00

Material cost per m² for a 3 coat wet dash in 3.5NHL   =   **€6.46**

Pointing

To analyse material quantities and labour costs for the pointing of rubble stone is difficult because stones and joints may vary greatly in size between one wall and another.

Take, for example, the amount of lineal metres of joints in 1 m² of rubble work which has stones not exceeding 200mm in height with some only 50mm in height (averaging c. 150mm) and lengths varying from 400mm down to 100mm (averaging c. 200-250mm).

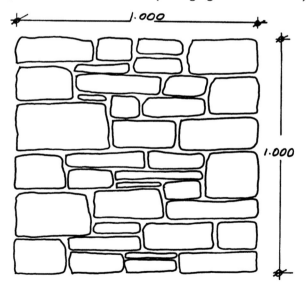

Estimating – pointing.

In such a case we may find that horizontal joints account for 9.50 metres and vertical joints 2.50 metres giving a total of 12 lineal metres of joints.

The average joint width is 20mm and the average depth 40mm.

The quantity of mortar required to fill these joints is thus 12 x 0.02 x 0.04 = 0.01 m³

It would be prudent on work like this to double this quantity to allow for extra width and depth of joints and waste. This will give an allowance of 0.02 m³ per m².

Using this figure we can now calculate the material cost to point 1 m² of rubble work which is similar to that explained.

Pointing - material costs

1:3 lime putty and sand @ €282.65 per m³ = €282.65 x 0.02 m³

= **€5.65** per m²

1:2 mix of 3.5NHL @ €223.26 per m³ = €223.26 x 0.02 m³

= **€4.46**   "   "

CHAPTER 11

THE GOBÁN SAOR

Stonemasons' legends

And now a little light relief! There are many legends in various countries about craftspeople who are held accountable for extraordinary building feats. In Ireland legend has it that the Gobán Saor is responsible for nearly every old stone building, in particular round towers and monasteries. The stories associated with the Gobán Saor were once well-known and handed down from generation to generation. The following are some of these stories.

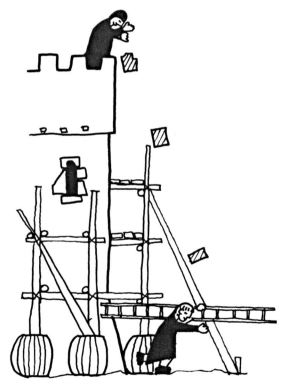

The Gobán Saor and renegotiation.

RE-NEGOTIATION

The Gobán Saor had just completed building a monastery in stone and asked for the monies agreed at the start of the job. The monks asked to re-negotiate this agreed price downwards.

The Gobán Saor would have none of this and refused point-blank. Unknown to him, as he was finishing off the top of the building, the monks removed the timber scaffolding and ladders. They then told him that they would not let him descend until he agreed to a lower price.

On hearing this he saw red and began to un-do his work, hurtling stones down from the top of the monastery, shouting that there were more ways to descend from a building than by a scaffold.

The monks became anxious and paid the agreed price!

THE CROOKED AND STRAIGHT TOOL

The Gobán Saor and his son left Ireland for seven years in order to build a castle for a king. Near the completion of the castle, the king decided to kill the Gobán Saor and his son so that they could not divulge any of the secrets incorporated into the structure of the castle.

The Gobán Saor suspected the king's intention and requested that he and his son be allowed to return home to Ireland in order to

The Gobán Saor in foreign lands.

bring back a special tool that was necessary to complete the castle. This tool was called a 'crooked and straight' tool. The king would not agree to this but suggested that his own son, the prince, would travel to Ireland to get the special tool.

The Gobán Saor agreed to this.

When the prince arrived in Ireland he asked the Gobán Saor's wife, Ruaidhseach, for the 'crooked and straight' tool.

On hearing this Ruaidhseach threw the Prince into a wooden chest and locked the lid. She knew that any request for a 'crooked and straight' tool meant that her husband and son were in danger. She then sent a message to the king that his son, the prince, would only be released if and when she saw her husband and son. The king agreed to the request and the Gobán Saor and his son returned safely to Ireland.

THE TWO-TAILED CAT

The Gobán Saor's mason's mark is the two-tailed cat. This can still be seen today on some old buildings.

The Gobán Saor was working at fifteenth-century Holy Cross Abbey in County Tipperary. The other stonemasons were jealous of his ability and did not invite him to dinner. He was told to find a two-tailed cat in order to be admitted.

He went outside and in a matter of minutes carved a two-tailed cat in stone.

He then fixed this in the wall and left. When the other stonemasons returned to work they saw the two-tailed cat but there was no sign of the Gobán Saor.

THE STRING LINE

The Gobán Saor, on completing the building of a stone wall, turned to his daughter and asked her if she had ever seen anything constructed so straight. She lifted up a piece of thread with which she was sewing and asked him if it was as straight as the piece of thread stretched between her two hands. He agreed his wall was not as straight. From that day forth he used a string line to build stone walls.

CHAPTER 12

LIME

Lime in Ireland is made from stone, unlike Holland who use shells. Preferably old stone from buildings and not fresh from the quarry. The manner of burning it into lime, usually all over Ireland, is this; in the side of some little height they make a great pit, round or square according as conveniently is offered; of that bigness as may hold 40 to 50 barrells, and of that fashion that being many feet wide at the top, it doth by degrees grow narrower towards the bottom as the same manner as the furnaces of the iron works. The inside of this pit they line round about with a wall built of lime and stone, at whose outside near the bottom a hole or door is left, by which to take out the ashes, and above an iron-gate is laid, which cometh close to the wall round about: upon this they lay a lay of limestone. (Being first knocked asunder with a great iron hammer, and broke into pieces of the bigness of a fist or thereabouts) and upon that a lay of wood or turf, or a certain sort of sea coal, the which being wonderful small, and peculiar called comb is hardly used for any other purpose. Upon that they lay another of limestone, and so by turns, until the whole kiln is filled, ever observing that the outmost lay be of wood, turf or comb, and not of limestone: which being done, the kiln is set afire until all is burnt.

Gerald Boate, The Natural History of Ireland, 1652.

Lime burning in Ireland, as described by Boate, continued for another 300 years and had probably changed little in the preceding centuries. From the seventeenth century lime burning became increasingly common in Ireland for land improvement. Lime is now recognised as a key element in the repair and conservation of old buildings. It was used by the stonemason, bricklayer, plasterer and painter. Unfortunately, the reference to using limestone from old buildings and not from quarries is met with in other old texts as well – obviously there was much unnecessary plunder of limestone from old buildings.

Lime was probably first introduced into Ireland during the period known as Early Christian (*c.* sixth – twelfth century AD). This period was characterised by Irish monks establishing monasteries right across Europe. We can assume that the monks became well-aquainted with what were essentially surviving Roman technologies, none more so than the use of lime in building. The Romans used lime for building brick and stone structures, and also for concrete, plastering, washes and internal decoration such as frescoes. For engineering works with water, such as sea piers and aqueducts, they added pozzolana to the mix to create hydraulic mortars and concretes. Pozzolana originated near Mount Vesuvius in Italy. One of the most famous architectural writers of all time, Marcus Vitruvius Pollio, a Roman architect in the first century BC, put all of this on paper in his *Ten Books On Architecture*. This book was a major influence later on renaissance Europe and is still worth reading today.

> There is also a kind of powder which from natural causes produces astonishing results. It is found in the neighbourhood of Baiae and in the country belonging to the towns round about Mt. Vesuvius. This substance, when mixed with lime and rubble, not only lends strength to buildings of other kinds, but even when piers of it are constructed in the sea, they set hard under water.
>
> Marcus Vitruvius Pollio, *The Ten Books of Architecture*, translated by Morris Hickey Morgan, Dover Publications.

Lime is obtained by calcining (burning in a kiln) limestone ($CaCO_3$) to a temperature of around 900° C. Sea shells (limpets, etc.) were also used for the same purpose in Ireland even into the twentieth century in areas near the sea where limestone was not readily available, such as in the granite regions of Connemara in the west and Tory Island off Donegal in the north-west.

In Ireland carboniferous limestone was laid down in tropical seas about 330 million years ago. This was a sediment of the remains of sea organisms, shells, etc. rich in the mineral calcite. Over time and with additional sediment laid overhead, the matrix was compressed by weight and bound together by calcite, forming hard, compacted, carboniferous limestone.

A newly constructed Welsh kiln built and fired at the Building Limes Forum (annual gathering), Wales 1997.

The sediments that make up limestone can be relatively pure in calcite and produce non-hydraulic limes when burnt; alternatively, they may be adulterated with mud containing minerals like alumina and silica, producing hydraulic limes. Non-hydraulic lime sets by carbonation while hydraulic lime has an ability or part ability to set in damp conditions, much like the pozzolana of the Romans discussed earlier.

Hydraulic limes, locally burnt, were used in Ireland and elsewhere in the past. Variations in limestone quality, firing materials and kiln temperatures produced lime that was classified into weakly, moderately and eminently hydraulic in the nineteenth century.

The only lime manufactured in Ireland today is non-hydraulic lime. Natural hydraulic lime is now being imported into Ireland from France and England as a hydrate in three grades: 2NHL, 3.5NHL and 5NHL. Use of finely ground, low-fired, clay-brick dust – a hydraulic additive for non-hydraulic lime and sand mixes – is also growing.

By burning limestone in a kiln, carbon dioxide (CO_2), which typically accounts for about 45% of the weight of limestone, is driven off. After burning we are left with quicklime, also called lump lime (CaO). Traditionally, quicklime was then mixed directly with sand, wetted and allowed sour out (see page 144), or the quicklime was slaked in water and run to lime putty ($Ca(OH)_2$) in a pit. This is discussed in detail later (see page 146). When non-hydraulic lime is used as a mortar, render, plaster or wash it sets by the re-introduction of carbon dioxide into the mix and the loss of water. This process is known as carbonation, and the cycle which began with limestone now ends as the carbonation process reverts the lime mix to stone (geological time-frame). Lime that sets like this is called non-hydraulic lime.

The lime cycle is summed up here as follows by an old Scottish stonemason in *A Treatise on Building in Water* by George Semple, Dublin 1776: 'When a hundred years are aft and gane then gude mortar is grown to a stain.'

Carbonation is most effective if lime is used in small overall amounts or thicknesses – a maximum coat thickness of 12mm in rendering or plastering, weak thin coats in limewashing and the overall reduction of lime mortar joints in rubble-walling by intro-

Lime kiln near the sea in County Wexford with old ship's timber as a lintel.

Circular lime kiln, pot-faced with conglomerate sandstone, County Wexford.

ducing stone pinnings into larger joints. Again this will all be discussed in detail later.

Traditional lime mortars seen in stone and brick walls were almost always hot lime mortars. Careful examination of old mortars may reveal that some are off-white – caused by unwashed pit sand with a high clay content, the introduction of mud into the mix, contamination of the lime with the firing materials, and so on. In some coal-producing areas a pink hue is evident from the anthracite culm used in the kiln which then finds its way into the mortar mix. Both the clinkering of the quicklime by over-burning and the mixing of the spent fuel with the mortar may have produced a hydraulic set in otherwise non-hydraulic mortars.

Mud was also used extensively as a bedding mortar in the past and once the building was rendered and limewashed it had an indefinite life span. At other times mud was used with stone in the core of walls while the outer leaves were built in lime mortar. The removal of renders, so popular today, but misguided, can expose mortars like these that were never intended to take the weather. Quicklime was also mixed with mud to produce mortars and renders.

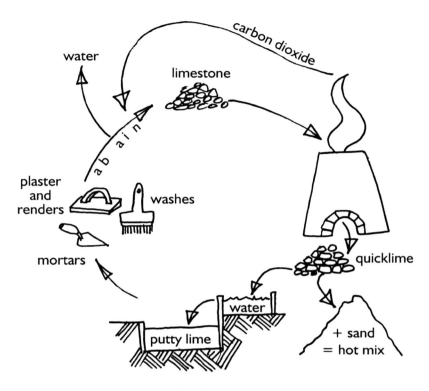

water

carbon dioxide

limestone

plaster and renders

washes

mortars

quicklime

water

putty lime

+ sand = hot mix

The lime cycle.

HOT LIME MORTARS

George Semple, in *A Treatise on Building in Water* (Dublin 1776), refers to his belief that ancient stonemasons working on churches, abbeys and castles were taught by Roman clergy to heart, or fill, walls with spalls and boiling grout or hot lime liquid so as to incorporate the spalls and the grout together like molten lead.

Modern hot lime mortars are produced by mixing quicklime with sand, adding water and mixing. In the past the quicklime was often buried in sand and let slake with rain. The result was high temperatures with steam rising from the sand. Normally this mortar would be allowed to 'sour out' or lie under cover in a wet state for six months or so with occasional turning over or mixing. After 'souring out' it was made ready for use by knocking it up. To knock up a mortar is to cut and beat it until it becomes plastic in consistency and therefore workable – a hard, strenuous activity. Hot lime mortars are recognisable by the small unslaked lumps of quicklime visible in the mortar. In mortars for building brick and stone these lumps simply acted as part of the aggregate, but in finished plasterwork they were susceptible to slaking any time they became wet or damp. The result of slaking is expansion and, therefore, popping of the plaster. Even so, hot lime mortars were used for scratch and float coats in plasterwork (but not finish coats) and for many external renders. Hot lime mortars produce wonderful 'sticky' mortars which bear no resemblance to modern mortars produced with sand and cement. They are a joy to work with. Sometimes hot lime mortars were used while still hot to kill frost or to accelerate the speed of drying in wet weather.

Hot lime mortars may look primitive in their production but the results produced are excellent.

Making hot lime mortars

This procedure should only be carried out by those trained in working with quicklime.

Safety must be a priority when working with quicklime – safety goggles, gloves, cover-alls, industrial footwear and head protection should always be worn. Close to hand there should be a plentiful supply of clean water as well as a medical eye wash. Quicklime reaches high temperatures well beyond boiling-point when in contact with water and will readily cause blindness and damage to the skin if there is contact. Any affected areas should be flushed with cold, fresh water immediately and, in the case of splashes to the eyes, hospital treatment is required immediately. A lime training workshop should be attended before engaging in working with hot lime mixes in any form.

Materials and equipment for working by hand (hot lime mix)

1. Fresh quicklime in sealed containers, preferably crushed but fist size will do.

2. Sand, generally coarse if for rubble stone. The sand should be well soaked with water first.

3 Hoe, rake, shovel, spade and pickaxe handle or similar.

4 Safety gear: goggles, gloves, cover-alls, industrial footwear and head protection.

5 A plentiful supply of clean water.

Making hot lime mixes by hand

But make the men temper it with the utmost expedition, and what you want in water to make it fit for your work, give it elbow grease; and this rule ought to be observed in making all sorts of mortar.

George Semple, *A Treatise on Building in Water*, Dublin 1776.

1. Introduce the quicklime into the wet sand and cover with sand as quickly as possible. Immediately there will be a thermic reaction and it is very important to cover the quicklime before it starts spitting, in case of splashing.

2. With the spade chop the quicklime lumps within the sand to reduce their size. If they become exposed keep covering them over with sand. Use the shovel to turn over the mix and the pickaxe handle to pulverise the quicklime lumps further.

3. Add water, working quickly because the quicklime will rapidly absorb the moisture in the sand and make mixing more difficult.

4. Continue chopping with the spade, turning over the mix with the shovel and beating with the end of the pickaxe handle. Add more water and continue mixing.

5. Steam is now rising from the mix and you should be perspiring unless you are superfit or it is freezing cold, in which case you will appreciate the heat rising from the mix.

6. When quite workable use the hoe and the rake. Keep working the mix back and forth until the quicklime is no longer visible except for tiny particles.

7. When thoroughly mixed, take a break – but not for long because the heat in the mix is evaporating the water and before long it will be difficult to work again. With practice, judging the amount of water necessary to create a final plastic (not dry) mix will become second nature.

8. When cool this mix should be soured out in containers

(plastic is fine if the mix has cooled) or stored in a large metal bin, on and under a plastic cover or similar. It is important both to prevent the entry of air and water loss. If frost is a possibility provide extra protection.

9. Knock up when required (traditionally after at least six months) by beating again, without resorting to the easy way out of adding water. Even though the mix may appear too dry, by beating it will become workable again. Adding water is the lazy way out unless a particularly wet mix is required as for wet dashing.

QUICK lime   sand   Water   souring out

Mechanical means to do the hot mixing and the knocking up can be employed, the best machine being the traditional roller pan mixer or mortar mill. This is a procedure best left to trained professionals because of the danger involved. Care must be taken not to stand too close to the pan in the early stages of the mix when the quick-lime may be reacting violently to the water and spitting out over the edge of the pan. The mill will crush the quicklime and physically work it thoroughly into the sand. All safety precautions as discussed before should be taken and a training workshop should be attended before commencing.

No machine will knock up the lime mortar better after being soured out than the mortar mill. It will usually do so without the addition of any water other than that already in the mortar.

The use of a standard cement mixer to make hot lime mortars is highly dangerous because the open drum, being nearly horizontal, will throw out boiling-hot materials at face level, a terrifying prospect.

LIME PUTTY

The running of quicklime to putty is best done by trained operatives because of the danger involved. Ready-made lime putty is available in tubs which eliminates most of the safety hazards of making your own.

Lime putty is produced by slaking quicklime. Quicklime is added to about twice its volume in water producing high temperatures, with boiling-point 100° C sometimes reached in less than a minute. The mixture far exceeds this temperature within a relatively short period. Lime putty, traditionally, was produced for plasterwork, particularly finish coats and decorative work.

The slaking-box was made of wood and sat at ground level.

During the slaking process in the timber box the quicklime is hoed or raked to assist it to break down so that it then can be run through a sluice gate with a sieve incorporated at one end of the slaking-box. The putty so produced is run into a pit. Traditionally, the pit was below ground level and lined with timber boards, or sometimes unlined. Excess water naturally drained away and after a period of time the putty thickened to the consistency of soft cheese. Lime putty is left in this state as long as possible – the older it is the better it becomes. Nowadays three to six months is common at times, but the Romans insisted in their specifications on at least three years. Non-hydraulic lime putty will only improve with age if it is kept moist and if carbon dioxide in the atmosphere is prevented from gaining access and causing a set. Preventing access by carbon dioxide is achieved by either maintaining a layer of water over the top of the pit, or by burying, or nowadays by using a plastic sheet. Alternatively, the lime putty can be stored in sealed plastic containers or plastic bags.

Making lime putty mortars

There is less excitement in producing cold mixes with lime putty but at least as much work.

This time we add no additional water, and we must use a dry or near-dry sand unless we are making wet dash. This is because lime putty has a high water content which is often enough to produce a workable plastic mix when beaten into dry sand.

Again, safety is important but we do not have the heat and thermic

Slaking quicklime to make putty.

reaction to contend with this time. Lime putty can still cause damage to the eyes and skin and, therefore, precautions must be taken as before.

Mixing lime putty mortars by hand

1. Put aside the water on top of the lime putty tub in another container. Spread half a 20 litre tub of lime putty along the base of the mixing trough and add sand (normally 60 litres if using a 1:3 mix). Then add remaining lime putty.

2. Using the hoe or rake, chop and mix the lime putty into the sand. This is hard work and even though there is no thermic reaction in the mix this time you may notice that you are undergoing a thermic reaction yourself!

3. Persevere without adding water. Use the pickaxe handle if necessary to beat in a vertical up-and-down motion.

4. Eventually the white lime putty will be dissipated throughout the mix and you will see a grey mortar. Keep going – you cannot overmix by hand.

5. Resist adding water. If water must be added use the water set aside from the top of the lime putty tub.

6. This mix is sometimes used right away, particularly if mature

lime putty has been used. Even so, it is considered better practice to sour out this mix and to store it away in a damp state for a period of time (minimum three weeks) without access to air or damage from frost. This mix is then knocked up when needed.

Natural Hydraulic Lime Mortars

Natural hydraulic lime (NHL) is a recent import into Ireland. We certainly had our own indigenous as well as imported hydraulic limes in the past, but have long since ceased to manufacture them. Hydraulic lime has many advantages when used externally, including resistance to frost because of a faster set, while still having elasticity. Hydraulic lime is available as a hydrate powder in sacks and can be mixed with sand and water in a standard cement mixer. It is classified as NHL2, NHL3.5 and NHL5. The choice of which grade to use is based on the nature of the host material (soft, friable, hard, impermeable), as well as environmental factors (marine/fresh water, high/low altitude, above/below ground, exposed/sheltered, time of year). As with all limes, it is not simply a question of getting the specification right – the skill of preparation, mixing, application and aftercare are at least as important. The following tables are based on St. Astiers Natural Hydraulic Lime.

Grade (N/mm² @ 28 days)	Historical classification	Frost protection	Kgs per m³	Masonry/ pointing NHL:sand	Sand – microns	Bag size (kgs)
NHL2	Feeble	96 hours	560	1:2	03:80	25
NHL3.5	Moderate	72 hours	600	1:2	04:80	25
NHL5	Eminent	48 hours	700	1:2	04:80	30

External Rendering	Scud (spatter-dash) c.1–3 mm NHL:sand	Scratch c.12 mm NHL:sand	Float c.10 mm min. NHL:sand	Finish c.5mm NHL:sand
NHL2	3:5	1:2	1:2	1:2/1:2.5
NHL3.5	4:5	1:2	1:2	1:2.5
NHL5	3.5:5.5	1:2	1:2	See NHL3.5 or NHL2

CHAPTER 13

POINTING

MICRO AND MACRO CLIMATES

Irish stone buildings are typified on their south and west elevations by weathered stone and mortar loss. These problems are created by the prevailing south-west winds which cross the country bringing rain. This rain at times can be horizontal and persistent, leading to rain penetration through the south and west façades of solid stone buildings. This can be most noticeable over windows, doors and reveals – in fact, anywhere the solid stone wall is punctured. Sometimes mysterious damp patches are to be seen on the internal faces of these walls, resulting from ingress of rain at a higher level. Effective solutions to these problems cannot always be found. Normally, wet and dry cycles dominate Irish weather patterns with relatively short periods of rain followed by dry spells. The majority of stone buildings are able to cope with this by quickly drying out if they have not been partially sealed by impervious pointing, renders and paints.

Different parts of buildings may suffer in different ways. For instance, chimneys are often subject to extreme weather condi-

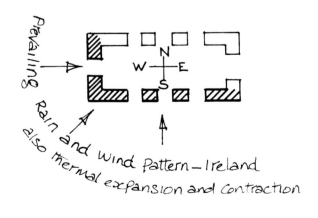

Prevailing Rain and wind pattern – Ireland
also thermal expansion and contraction

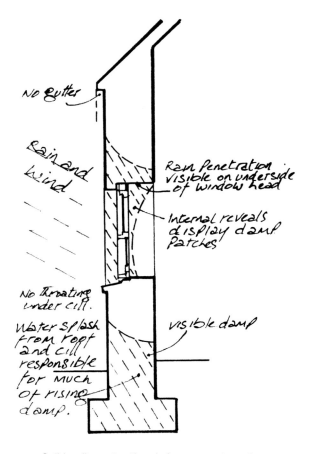

Solid wall construction during very wet weather.

tions, being cold externally while at the same time warm internally. The action of flue gases, soot and water cause sulphates which expand and burst stone and brick, and even cause 'jacking' of the chimney off-plumb.

The south and west faces of buildings are also subject to extremes of expansion and contraction from the heat of the sun followed by shade, or warm days followed by cold nights. This is counteracted by the elasticity of lime-based materials in traditional buildings.

A building requires pointing when there is

- Excess mortar loss from joints allowing ingress of rain, growth of plants, rounding of stone arrises from weather or similar damage. Open mortar joints, even if viewed from a distance, nearly always look darker because of the shadows thrown. Mortar loss may be so extreme at times that instead of pointing, which is the replacement of surface mortars in joints, it would be more correct to use the term replacement of bedding mortars.
- Loss of stone pinnings from joints in rubble work, exposing large areas of lime mortar in recessed pockets. Larger stones on the face as a result are becoming loose and there may be partial collapse of the external stone face. The loss or near loss of stone pinnings is, in effect, the building's cry for help, by pointing.
- Previous pointing has failed and needs to be replaced or is causing problems by being impervious and accelerating decay of stone and preventing the wall drying out.
- Existing mortars have failed, dis-aggregated, and are capable of being brushed out.
- Pointing should be done in advance of rendering or plastering, if deemed necessary.

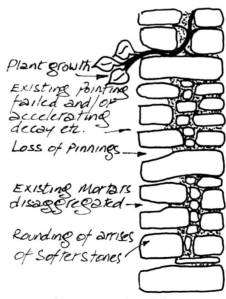

Some reasons for pointing.

A building does not require to be pointed for any of the following reasons

- For aesthetic purposes: for instance, raking out and pointing perfectly sound areas of original mortar just to match in with an area that had to be pointed. Global pointing of buildings is rarely necessary and selective pointing as necessary is best. This may result in a difference of colour initially between the old and the new but in time this will even itself out if the appropriate mix and technique have been used on the new pointing.
- If existing mortars are relatively soft and capable of being easily raked out. This is particularly so where there are no obvious problems such as the ingress of rain as a result. If these mortars are relatively flush to the face of the stone leave them alone.
- Because it is generally felt to be a good thing – and 'don't all buildings need to be pointed?'

A serious need for pointing, County Louth.

Why original bedding mortars fail

- Severe weathering, such as on parapets subject to high winds, rain and weathering from both sides. Also, leaking gutters and downpipes concentrating water on specific areas. Water spray or splash from passing traffic on the lower section of walls.
- The quality of the lime and/or sand was poor. Lime may have been air slaked before use, under or overburnt in the kiln, not mixed thoroughly enough with the sand, etc. Sand may be poorly graded or have high

concentrations of clay or other impurities.

- The original lime and sand mix ratios were either too rich or too weak. If too rich they may have caused excess shrinkage and cracking. If too weak, insufficient lime was present in the mix to bind the sand together.

- Insufficient protection was given to the work immediately after laying from rain, wind and sun. This resulted in the loss of mortars near the face of the wall.

- Non-hydraulic mortar was used where a hydraulic mortar was needed, such as in continually damp areas.

Severe weathering with mortar loss and 'rounding' of softer stone, County Louth.

- The necessary protection of an external render is missing, having been removed or simply weathered away over time, leaving bedding mortars exposed to the elements. These mortars are now in turn failing.

- Large, exposed, wide joints in rubble work were insufficiently pinned with smaller stones.

- Ashlar work with tight joints laid in pure lime putty without a fine sand has resulted in joints turning to powder and being washed out.

- Ingress of water, particularly through the tops of walls over time, may have washed out lime mortar cores in the centre of the wall with the danger of the remaining core slipping and causing structural failure to the wall. These core mortars will require replacement by low-pressure or gravity-feed pumps. This in itself has to be controlled as the introduction of a large fluid mass into the heart of a wall may cause more problems than it cures.

Why previous pointing mortars fail

- They were applied over flush or near-flush original bedding mortars. In other words, they were not required in the first place and secondly, either no raking out occurred or it was done to an insufficient depth.
- Very hard, dense, impervious pointing mortars such as rich mixes of sand and cement were used which cracked and pulled away from backgrounds, allowing water in behind but slowing down evaporation. Moss/algaes can be seen in behind these pointing mortars where it is naturally damp. Frost will cause a certain amount of damage.

What constitutes good pointing and replacement mortars?

- Sacrificial mortars – they will fail sooner than their host material, brick and/or stone.
- Pervious mortars – they do not prevent wet walls from drying out.
- Matching as far as possible existing mortars in the wall in terms of aggregate size, colour, shape, geology, binder, aggregate ratios, etc.
- Non-permanency – capable of being removed without damage to the surrounding stone or brick.
- Designed to suit their geographical location. Severe weather exposure in areas such as mountains will subject mortars to longer periods of rain and freezing than most low-lying areas.
- Designed to suit their location on the building – wall tops, barges, sills, continually damp areas, those with limited access to carbon dioxide, piers or locations which take tensile strain require different mortars than flat, vertical areas of walling.
- Designed to suit their host material which can vary from being soft and friable to hard, dense and nearly impervious stones or bricks.
- Designed to suit their joint size and depth.

POINTING PROCEDURE FOR RUBBLE WORK

Raking out

1. Rake out on average to a depth of twice the joint width with the proviso that common sense be applied to this rule in case face stones become dislodged where joint widths are large. A minimum depth should also be applied of say 40mm.

2. Raking out should commence at the top of a wall and move downwards.

3. Any stone pinnings which are loose or become dislodged in raking out should be re-inserted dry into the joint until the pointing is commenced when they can be properly installed in their original position.

4. Hand tools such as plugging chisels and lump hammers should be used for raking out.

5. It is important that the top and bottom arrises of each joint are cut out square and clean – simple V cuts are not acceptable.

6. Raking out should be inspected before pointing is commenced to check for depth, cleanliness and so on.

7. Brush down work, making it free of dust and debris.

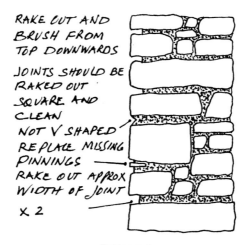

Raking out.

POINTING

1. Joints should be dampened down in advance of pointing. This may involve a simple, plastic, hand-held spray-bottle filled with clean water or a hosepipe with an on/off spray connection. Whichever is used depends on the joint size, area of work, soakage rate of background material and size of work force. It is critical to dampen joints in advance of pointing but this can be done insufficiently or excessively. If done insufficiently pointing mortars may lose their water content too fast from background soakage and fail as a result by powdering or cracking. Excess wetting of the background may introduce water internally into the building, or leave work liable to frost damage. Every operative engaged in pointing should be supplied with a water-spray bottle to keep joints dampened in advance. Depending on the time of year and other factors it may be best to give work a good soaking and allow to dry until damp before pointing.

2. Mortars for pointing should be well-worked and beaten but have a low water content. Less water in the mix results in less shrinkage and also enables mortars to be compressed efficiently into place. Faster setting times will also result with lime-based mixtures as they must lose water in order to set: this can be slow in cool, damp weather with stones having low soakage. Low water content also means less possibility of staining to stone faces than would occur with wetter mixes, and less chance of frost attack.

3. Application of pointing mortars should be with a flat steel (or non-ferrous) bar. Each operative needs to be supplied with a range of different widths, generally 6, 10, 12, 16, 20, 25, and a 30/40mm width. A flat bar that is too narrow will not properly compress a wide mortar joint and one that is too wide will be ineffective on smaller joints. Most of these bars are best made from flat steel bars supplied by a steel supply works. They are relatively inexpensive. Most shop-

bought varieties are too flexible because they are designed for modern, soft, plastic mixes on new work – the range of sizes is not available either. The length of the flat bar is important also and on work which exceeds just pointing and includes replacement of bedding mortars quite long examples may be required.

Flat steel or non-ferrous bars.

4. Mortar is applied into the pre-dampened joint and brought out flush to the face of the wall. Large joints will have stone pinnings knocked in with a lump hammer using light taps. Horizontal pinnings are tilted slightly outwards. An average joint width not exceeding 20mm around the pinning should be sought. On accurately-controlled work, by the insertion of pinnings, an average joint size of 10-12mm is possible, and as a consequence work which would otherwise look mediocre visually springs into life. On the replacement of historic mortars it would not be appropriate to introduce pinnings into joints where there is no evidence that they ever existed. Care should be taken not to stain the face of the stone with mortar. A hawk with a projecting sheet-metal top can be useful for this purpose. Very deep joints are sometimes filled in stages.

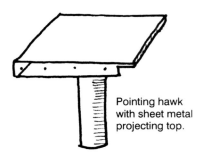

Pointing hawk with sheet metal projecting top.

5. The flush joints when quite hard but still slightly plastic are then beaten with a stiff nylon brush. This closes any minor shrinkage cracks and also exposes the aggregate in the mortar, which helps it match in with older work if the correct aggregates and binders have been used in the new mix. A uniform even texture is also obtained, giving the work a finished, professional look. Lime and sand mixes can some-

'Beating' brush.

times be beaten at the end of the day's work or if the weather is cold and damp it may take longer before the work is beaten. To beat the mortars too early will result in the marks of the nylon brush being left on the face of the mortar joint and the aggregate will not be exposed. Care is needed and timing is critical. On a large job an operative could be assigned full-time to this work. Other areas that can be assigned within work teams are the pre-wetting of work in advance of others and the light spraying of work which is drying too fast. All lime mortars benefit from an occasional fine mist spray during the week after placement.

6. At all stages work must be protected from rain, frost and sun. At night if there is any danger of frost the work needs to be adequately protected by plastic sheeting, hessian and possibly tarpaulins. At times polystyrene boards may have to be used underneath a cover but these are so light it is hard to keep them in position. The aftercare and protection of work is critical. Scaffolding, in city areas in particular, may be covered with clear plastic sheeting which is reinforced. This is used to protect the public from dust and debris but also can protect work from driving rain.

Incorrect styles and materials for pointing

- Strap pointing has no place on traditional stonework and should not be used on modern work either. Basically it is a projecting strap of cement mortar laid over joints, great and small, and cut to an even width. Small joints are hidden in behind it and wider joints are camouflaged either side. Unfortunately, it is very popular not only in Ireland but many other countries as well. Attempts to remove it can be nearly impossible if the cement-to-sand ratio in the mix is high, which it usually is. If capable of being removed it will leave a near-permanent stain on the face of the stone. It causes problems, like all impervious pointing, by preventing the wall from drying out when wet and can accelerate decay in softer stones by concentrating areas of dampness around joints which are then subject to frost and other forces.

- Weather struck pointing is also very popular but again, not

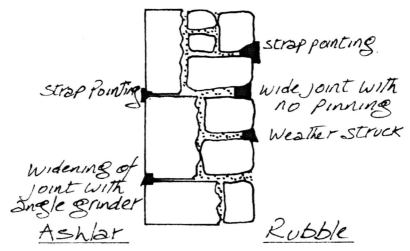

Wrong methods and styles of pointing.

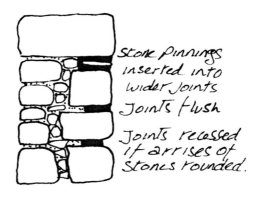

Correct styles of pointing.

appropriate on stone. The idea behind sloping joints is to shed rain like roof slates. Normally executed in sand and cement it has no place on stone walls.

■ The application of a wider joint over ashlar joints is to be con-demned for the same reason as in strap pointing. Additionally, it makes otherwise finely cut and laid stone look coarse. One of the reasons it occurs is that the maximum aggregate size of the sand used is too large for such small joints and instead of working with a finer sand the locally available sands are used and are unsuitable. In most rubble work these sands are too fine but in this work they are too coarse.

- The cutting out of fine ashlar joints with cutting discs in order to make them wider and easier should be punished by hefty fines. It is quite common, and results in the permanent destruction of fine work. Joints once 3mm and less are converted to joints of c.12mm but, worse still, are over-run into adjacent stones and are rarely straight but skewed or in wave patterns.

- Large flush joints in rubble work with no pinning stones and observable brush marks. Again, very common on both old and new work. Concentration is only on the large stones and smaller stones are forgotten. Cement mortars allow this to happen because of their faster setting times and their ability to set even in large areas/volumes.

ASHLAR POINTING

Fine-jointed ashlar work is often unnecessarily raked out and pointed with the wrong materials, styles and procedures. It is probably the most difficult type of pointing except for the tuck pointing of brickwork. Inevitably, an effective approach is developed for each situation and what works on one site is only partially successful on the next. For this reason be prepared to try various non-destructive methods using different tools until you find a way suitable for your situation. This is actually true of all work described in this book.

Cornices, string courses, sills and barge stones are often in ashlar and lose their vertical mortars not because they have thin joints but because their joints were filled in-situ after they were laid. This is never as successful as applying cross joints in advance and squeezing the next stone into position thereby compressing the mortar. Non-hydraulic mortars in these areas, sometimes with little or no sand, have a short life expectancy.

When replacing missing mortars in these areas the new mortar should preferably be hydraulic and be compressed into position with specially-made flat steel bars.

Raking out of ashlar joints

1. Industrial hacksaw blades, thinly tapered chisels, thinly ground plugging chisels and basically any hand tool that will take out mortar without causing damage to the arrises of the stone or widen the existing joint size can be used. Difficulty is met sometimes, not with taking out the existing bedding mortar but with undoing previous hard sand and cement pointing. To remove pointing like this (which is rarely deep) it is necessary to fracture its face by using a small sharp point or a thinly tapered sharp chisel held at near right angles to the work and struck. At times existing mortar loss is so severe on ashlar work, particularly on the south and west faces of buildings, that little raking out has to be done, but bedding mortars do sometimes have to be replaced to excessive depths.

2. Do not use angle grinders and other blade-cutting equipment which cause much damage by widening joints, over-running into adjacant stones and causing uneven joint shapes.

3. The rake out depth should not be less than 20mm and should be 25mm if possible. The joints should be raked out square and clean.

4. Work should be brushed down thoroughly of all debris.

Pointing ashlar work

1. Pre-wet joints in advance of pointing with a spray-bottle having a controllable nozzle capable of being directed at the joint. A syringe is also very useful. Allow for joints to dry until damp.

2. Mortar should be placed in position using a specially-made flat jointing iron that is capable of being inserted into the joint. Steel flat bars are readily available in different thicknesses and widths such as 3mm x 10mm and 5mm x 12mm. Normally the largest dimension and the width of the flat bar is used to insert mortar, but in the case of ashlar work the edge side is used. The bar can be twisted halfway along its length as illustrated (see page 157) to make this easier.

3. A hawk with a projecting sheet-metal top is handy at times

for protecting the work underneath and for getting mortar close to the workface.

4. Masking-tape can be applied over the joint and split with a sharp knife so that mortar can be inserted into the joint without staining the stone.

5. The mortar is pressed firmly home and brought out flush to the face.

6. The finished joint can be tamped flush with a brush. This is particularly effective if there are existing broken arrises on the stone and the mortar chosen is similar in colour to the stone – it will make the damage on the stone less obvious. Sometimes joint faces are left smooth from the flat bar and flush to the face of the stone. The pointing should not project past the face of the stone.

7. Immediate aftercare is critical. With such a small amount of mortar the stone may very quickly absorb its moisture content and cause accelerated drying out resulting in failure. Control can be maintained by protecting with plastic sheeting until set.

CHAPTER 14

PLASTERING

Internal plastering is traditionally done on backgrounds of wood lath, stone and brick.

The composition or ratios of mixes varies little on these backgrounds but their suction rates can affect drying times, either making them dry very slowly or too fast. Work that dries too slowly will delay production and increase costs while work that dries too quickly will crack and fail. Backgrounds with a high suction rate such as brick can be pre-wetted using a weak limewash (1 lime putty : 9 water) to control the rate of drying; other backgrounds, such as a hard, dense stone in cool, damp weather, may require very little pre-wetting.

Most non-hydraulic lime putty: sand mixes are c. 1:3. This is based on sand having voids (air) of 33%. The lime putty, therefore, simply replaces the air in the sand. Where the voids exceed 33% by volume additional lime is added accordingly.

Void area can be measured by a simple field test using dry sand and water.

COMMON MIXES

Pointing (internally)

> 1 lime putty to 3 sand (1:3) with hair at 3-5kgs per m³
> plus stone or brick pinnings as necessary.

Daubing out

> 1:3 with hair as before in multiple coats of max. thickness 12mm each plus stone or brick spalls on flat if necessary.

Scratch coat

> 1 putty to 3 sand (1:3) with hair as before, max. thickness 12mm

Float coat

I putty to 3 sand (1:3) with hair as before, thickness 6mm – 8mm

Finish coat

I lime putty to I fine sand (1:1) with no hair, average thickness 3mm

Hair

Traditionally the hair used in plasterwork was ox hair, often reddish in colour. Cattle are no longer wintered outside and as a result do not grow hair as a protection against the cold. However, goat and yak hair is imported for plasterwork. Animal hair must be obtained from a reliable source to ensure that it is clean and free from disease. Goat's hair averages between 75mm and 150mm in length while yak hair can be 400mm long. The optimum length for plasterwork is between 25mm and 75mm in length. Hair should be teased into the mix so as to be distributed evenly and not in clumps.

POINTING (INTERNALLY)

Pointing is sometimes necessary internally before plastering begins. It should only be done selectively (where needed) with lime mortar and finished flush to the face of the wall with a coarse finish. Stone or brick pinnings should be inserted into larger joints. The mortar used should be stiff and inserted into clean, pre-wetted (lime-washed) joints with a steel flat bar. Haired-mortar, as used on the scratch and float coats, is ideal.

Daubing out

Daubing out – required when excessive hollow spots occur on the face of the wall – should be done in multiple coats not exceeding 12mm each in thickness, to allow drying out and partial carbonation to take place before adding subsequent coats. Each coat should be finished with a wooden float and scratched to receive the next coat. Where excessive depths are met with in daubing out (over 25mm), flat brick or stone spalls should be used to reduce the overall volume of mortar used in the daubing out. Thin slivers of porous clay brick are best for this as they accelerate both the drying out

Tools and materials used for plastering.

and carbonation processes. They should be pre-wetted before use. Pre-wetting (limewashing) of background surfaces and between coats is also usually necessary.

Scratch coat

The scratch coat is applied next to a maximum thickness of 12mm and finished with a wood float.

The work should be checked for shrinkage cracks and if they occur should be closed down with the wood float. Shrinkage may appear a few days after application depending on the weather. It is critical that the next coat is not applied until all risk of shrinkage is eliminated. In general once shrinkage cracks are closed down they do not return again. Work applied on pre-wetted (limewashed) backgrounds in average weather conditions rarely shrinks.

Traditionally, scratching was done using a number of timber plastering laths nailed together to scratch the work diagonally. Diagonal scratching was important on timber lath backgrounds because of the danger of puncturing the scratch coat through the horizontal gaps between the laths. The same tradition of diagonal scratching was carried over to brick and stone backgrounds. Scratching is done in order to provide a key for the subsequent coat but it also achieves faster setting times by allowing carbon dioxide greater access to the lime/sand coat. When the scratch coat is is no longer liable to shrink it is ready for the float coat. A practical test is to try to make an impression with the knuckle of one's hand – it should be barely possible.

Float coat

The float coat is applied to a thickness of 6-8mm and finished with a wooden float. The surface is then lightly scratched using a devil float, which has small nails that project about 1mm. The scratch coat should be pre-wetted (limewashed) first to prevent the float coat drying too fast. Check for shrinkage and close down with the wood float as necessary.

Finish coat

The finish coat is applied to a thickness of about 3mm and finished with a steel trowel. The float coat should be pre-wetted (limewashed) first. Pre-wetting is critical or the finish coat, because of its relative thinness, will lose its moisture rapidly to the background coats if they are dry. Aftercare is critical when working with lime because it has slower setting times than cement or gypsum mixes. Lime mixes are vulnerable during their drying and initial carbonation periods – it is risky to apply a finish coat on a Friday afternoon when nobody will return to work until Monday unless the plasterers are well experienced with the materials, background, soakage rates and prevailing weather.

In the past, plaster of Paris was gauged (mixed) in with the finish coat. This was used where finished work was vulnerable, such as at corners, although timber beading of various widths was also used in these locations. The nearest equivalent today to plaster of Paris is gypsum wall finish, added if

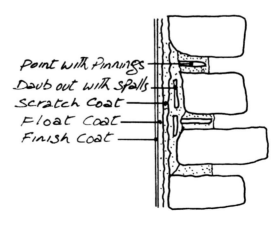

Internal plastering.

necessary to the finish coat at about 25% by volume.

Corners were sometimes rounded to prevent damage and this can nowadays be an essential repair procedure to maintain the character of the original work.

Sand

The sand in all mixes should be sharp, clean and well-graded. For the pointing, daubing out, scratch and float coats, use the same sand with a maximum particle size of 3mm. A sand with a maximum particle size of 5mm is best for pointing, daubing and scratch coating only; this is a little too coarse for the relatively thin float coat of c. 8mm.

The sand for the finish coat should have a maximum particle size of 1mm (B.S. 1.18mm) and be sharp, clean and well-graded. Good quality fine sand is difficult to obtain and for small works it is best sieved from the previous well-graded sand. This entails running the sand dry through a 1mm mesh. Some of the finer sands available need to be checked for grading – sometimes they are of a single grade with no fines (fine grains), which means they have large void spaces resulting in shrinkage when used.

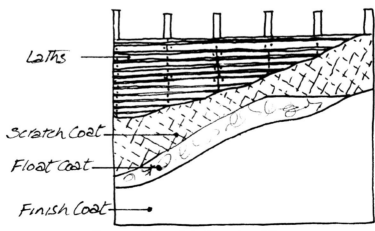

Three-coat internal plaster work on laths.

A lime and sand scratch coat with animal hair being laid over timber laths.

Plaster laths

Plaster laths in the past were riven, or in other words, split along the grain of the wood. This resulted in a lath of variable thickness along its length. They were also twisted and varied in width. Because they were split on the grain they were both supple and strong. They were fixed with gaps of 6mm to 10mm, average 8mm, between each, to timber studs and joists using iron nails. The iron nails which we find in older work today are always rusted but give few problems as a result except, occasionally, on ceilings. Lath widths on older work were at times excessive, resulting in fewer keys to fix the scratch coat and occasional failure as a result.

Machine-cut laths can, of course, be used today and were used occasionally from the second half of the nineteenth century onwards. For partitions an average size of lath is about 25mm x 6mm and for ceilings 25mm x 10mm. The thicker lath is necessary on the ceiling to carry the weight. Gaps between laths are about 8mm. Studs are traditionally fixed at around 300mm centres.

Today, galvanised clout nails should be used in repairs. If there is any danger of causing damage to otherwise sound plasterwork on ceilings and partitions by nailing new laths into position then they should be fixed with non-ferrous screws.

Laths were used all over Europe, North America, Australia and else-where until well into the twentieth century prior to gypsum sheet materials.

We have looked only at three-coat work, ie, scratch, float and fin-ish, but, of course, one- or two-coat work was used sometimes on poorer grade work.

All coats benefit from a regular mist spray to assist carbonation and reduce shrinkage. Hydraulic limes may be necessary on stone and brick walls internally, in basements for instance.

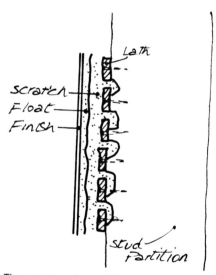

Three-coat work on a lathed partition.

EXTERNAL RENDERING

In Ireland practically every traditional stone building – farmhouse, townhouse or shop – was rendered. Court houses, churches, schools and large houses on estates were built in stone which was meant to be seen, but even they were sometimes rendered. The current fashion of removing renders is to be condemned as it is changing the original visual and aesthetic character of Irish towns, villages and individual houses. It also creates problems with rain penetration, heat loss, deterioration of timber lintels, internal plasterwork and paint.

Decorative external mould details – barbarously removed today from around windows and doors – allowed these buildings to be dated and often reflected a local style or tradition that was handed down from father to son. Some owners of old buildings have had to make good the loss by re-rendering, incurring double the expense.

It is quite obvious that some stone faces, particularly when relatively smooth, were 'sparrow pecked' with a punch in advance of being rendered. This is visually different from normal punch work patterns. The bottom arrises of stones were sometimes kept in from the face so as to create a stepped pattern on the face of the wall on which to hang a render. This encourages ingress of water when renders are removed.

Worse still, many of these stone buildings are pointed with sand and cement which prevents the structure drying out after rain.

The materials used in external renders varied greatly compared to internal plasters. Hydraulic limes when available were used in the past, including brick dust to give a hydraulic set, but also glass, coal, etc. Patented cements such as Roman cement (manufactured in Britain in the nineteenth century) were used and when Portland cement became available it was first used in external renders. Other renders incorporated oil and mastic. Traditional wet dash or harling could be any of the varieties of hydraulic or non-hydraulic limes.

Quoin stones imitated in stucco, Parco Demidoff, Pratolino, Florence, Italy.

Some external renders were too rigid, causing severe cracking or crazing and, at times, pulling away from their backgrounds.

Renders were sometimes ruled to imitate stone and were called 'ruled ashlar' work. These painted renders are still common in villages and towns. Materials thrown on as an external render are referred to as wet dash in Ireland, harling in Scotland and roughcast in southern England.

Wet dash

There is a great European tradition of throwing mortars at walls, either leaving them as is or working them immediately with wood floats and trowels. In some countries the lime mortars were beaten before setting on the wall. All of this makes practical sense – thrown-on mixes have greater surface areas and therefore dry and carbonate more successfully. Beaten renders will lose their water content quicker and therefore dry and carbonate quicker also. Practice always make sense no matter how strange it seems.

Wet dash can be found on houses, barns, churches, boundary walls and a multitude of other structures. In general, the material used as the wet dash is identical to the bedding mortars in the wall itself. Some wet dashes are quite thick while others are extremely thin. The thick coats are often multiple coats but, occasionally, single coats were thrown on with stiff material. The thin coats are mostly the result of wet single coats thrown on and then weathered back.

Old, weathered wet dashes are hard to match because weathering exposes the aggregate in the mix and the surface appearance is rather flat. New wet dashes, on the other hand, have their aggregates covered in lime and look a little sharp.

Compared to the large even particle size of modern wet dashes traditional dashes are generally finer – maximum particle size of around 6mm to 8mm is common. These aggregates are, of course, graded from a maximum particle size down to dust. An analysis of the aggregate size, colour, shape and its geological type can be made if necessary. The amount of lime in the original mix can also be determined.

Wet dashes which have been limewashed with or without a colour in multiple coats have a soft, rounded, textured look which is quite beautiful.

New wet dash (lime and sand), later limewashed, County Kildare.

How to wet dash

To wet dash a wall, pointing and daubing out, if deemed necessary, should be taken care of first. Pointing needs to be done with care, remembering to use stone pinnings in large joints (see Chapter 13, Pointing). Daubing out is the filling out of hollow spots on the wall and this should be done in coats not exceeding 12mm in thickness. Stone, or better still, flat brick pieces can be used in the daubing out to reduce the overall thickness of mortar in the daubing.

A wet dash can be thrown on as a single finished coat or in multiple coats. Background coats can be pressed back with a wood float if necessary but the final coat once thrown should be left as is. It was not normal in the past to lay on a scratch/float coat prior to wet dashing (this is a modern practice).

An application of only one coat will in most cases leave the shapes of stone and mortar joints still visible through the dash to an excessive extent and, therefore, up to three coats are normally applied.

Mixes vary from c.1 putty lime:3 sand to 1 hydraulic lime:2 sand. For most conditions over hard stone a hydraulic lime is preferable. Hybrid mixes using hydraulic lime, non-hydraulic lime and sand can also be used with proper advice and supervision.

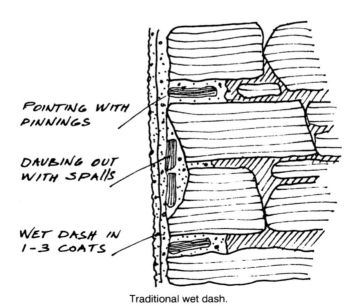

POINTING WITH PINNINGS

DAUBING OUT WITH SPALLS

WET DASH IN 1-3 COATS

Traditional wet dash.

Limewashing should be seen as an essential part of non-hydraulic renders such as wet dashing.

Background surfaces should be pre-wetted with a weak limewash.

Ruled ashlar

Ruled ashlar imitates finely cut ashlar stone. The sub-strata behind these renders is sometimes quite crude rubble stone. Again it is worth stating that these renders should not be removed to expose the stone. Ruled ashlar and other renders are commonly painted. In towns and villages individual houses and premises are often of different colours which can be lively and interesting.

Renders like ruled ashlar are normally finished to a flat plane surface with a wood float. Traditionally, to achieve a very flat overall appearance, boards or edges were fixed vertically at each end of the building, projecting out far enough from the face of the building to avoid any high points on the face of the wall. These boards were

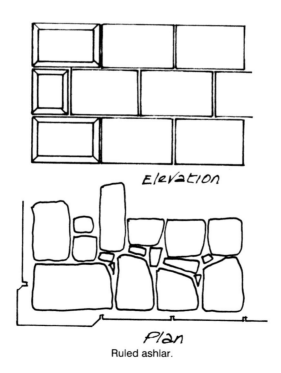

Elevation

Plan

Ruled ashlar.

then plumbed and/or sighted with each other and a horizontal string line was pulled between them at about 1.8 metre intervals measured vertically. Nails were driven in to the wall until the nail heads were flush with the line. The nails were fixed at about 1.8 metre centres horizontally.

The string lines were removed and vertical mortar screeds applied flush to the nail heads. The areas between the screeds were now filled in with two-coat work until flush with the screeds.

Finally, the last coat was applied over the previous two and their screeds. This was ruled with a jointer while still soft, both with horizontal and vertical joints, to give the appearance of stone. Horizontal joints were marked out on the end boards first and string lines drawn across the face of the building. The render was marked to the line in places so that a straight edge could then be applied to the render in order to run a jointer to mark out the joints on the work. It was important here not to leave ragged edges on the joints by using too large an aggregate size in the final coat or the wrong jointing tool.

External renders, traditionally, are laid on to a wall in two or three coats working out from strong coats to weak coats, ie, the final or finish coat is the weakest mix. This rule is still applied today on modern external renders.

Hydraulic mortars were commonly used for smooth renders like ruled ashlar. Roman cement imported from Britain, hydraulic lime and Portland cement were all used. Smooth renders in non-hydraulic lime only mixes carbonate slower than the wet dash variety because of a smaller surface area, and are not as successful.

CHAPTER 16

LIMEWASHING

The tradition of limewashing buildings is to be found in many parts of the world such as Greece, Italy, Portugal, Spain, Britain and Ireland and wonderful coloured limewashes are still to be seen in Scandinavian countries, Russia, North Africa and elsewhere.

Limewashes not only gave life and colour to old buildings but were used as an antiseptic against disease – to this day in Ireland it is used for this purpose in cow byres.

Many countries use pictures of a limewashed vernacular house to attract tourists. In Ireland this is typically a white, limewashed cottage with a thatched roof and strongly coloured windows and doors. The first of May was a favourite time for limewashing in Ireland. Limewash was used externally and internally on vernacular buildings. From the eighteenth century distempers were used internally on larger buildings.

Limewashes have a distinctive look which is unmatched in any other paint medium.

Some old buildings show evidence of multiple coats of limewash applied over the years, sometimes dozens. The study of these coats to reveal the colours used is important for any future conservation and repair work. Often the first intervention made by a new owner of such buildings is to remove these washes and the underlying renders and plasters and to clean the walls back to their stone skeleton, which is most unfortunate.

Limewash is like a protective skin that allows solid walls to breathe, a critical issue in traditional solid wall construction as discussed before. Modern impermeable plastic paints impede drying. Walls stay wet longer, rising damp climbs to higher levels and when dry, salts expanding behind the plastic paint lift it off in sheets sometimes including the renders and plaster. Many old buildings have peeling plastic paint internally but to remove what remains can be most difficult and can damage underlying plasterwork. Again, the

A Venetian red finish, Pisa, Italy.

Vernacular cottage displaying the remnants of various limewashes, County Galway.

first principle of conservation is that anything done should be capable of being undone.

The limewashing of lime plasters and renders has the additional benefit of introducing lime-rich water which, on carbonating, fixes or re-fixes them back to their sub-strata. Small cracks in the render are also closed. New renders to be limewashed must be permeable – an impermeable render results in the mix simply lying on the face of the render and being washed off with the first shower of rain.

How to limewash

On new work external limewash is generally applied in five coats with mixes as follows:

- Coats one and two I part lime putty to 9 parts water (1:9)
- Coats three, four and five I part lime putty to 6 parts water (1:6)

If the desired effect is not achieved additional coats of 1:6 are applied.

Mixes can vary somewhat to achieve the desired effect but multiple thin coats are preferable to thick coats which shrink, crack and powder.

Use lime putty rather than hydrated bagged lime as it carbonates more efficiently and is less likely to crack and wipe off on clothes.

Externally, the final coat has the addition of a fixer such as tallow (clarified animal fat) or raw linseed oil. These materials are added at about 1/50 of a part to the limewash. Internally, three coats are normally sufficient and no fixer is added. The fixer ensures that the external limewash will last a reasonable amount of time in average conditions.

The heat generated by slaking quicklime and water is sometimes made use of when adding materials like tallow and raw linseed oil to increase their dispersal in the mix. As this is a dangerous activity it is more normal to use cold lime putty instead with raw linseed oil thoroughly stirred into the mix.

The following earth pigments may be added to all coats or just the final ones depending on the finish required. Very interesting finishes can be achieved by lightly laying a coloured limewash over a white limewash, or one colour over another, such as venetian red over raw sienna.

Pigments

- Venetian red gives a colour varying from light pink to dark red depending on how much is used.
- Raw sienna, a light brown/yellow ochre colour.
- Burnt sienna, a warm brown colour.
- Yellow ochre, a weak yellow colour.
- Copperas wash using ferrous sulphate (see below).

Any pigment used must be lime-friendly.

Pigments can be mixed with warm water by shaking in a sealed glass jar before introducing to the limewash to disperse the pigment thoroughly through the mix. If pigments are added directly to the limewash they need to be thoroughly mixed in or slight streaking may occur as undissolved pigment is dragged across the wall surface. This, on the other hand, can be quite attractive.

A copperas wash was common in nineteenth-century Ireland, the remains of which can be seen on many old buildings. Also used in

Scandinavian countries and elsewhere, the term copperas is confusing as the material used was ferrous sulphate (green crystals). This wash goes on green but quickly changes to a yellow-butter-tobacco colour depending on the amount used and the background surface.

Ferrous sulphate is first dissolved in warm water in a glass jar as before and then added to the limewash. It should not be used on internal wall surfaces.

Tools and equipment

- A wide distemper/emulsion brush
- Lime putty
- Clean water
- Empty buckets
- Earth pigments as required
- A hand-held egg-beater or a whisk attachment for an electric drill (used with care)
- Safety goggles to protect eyes
- Gloves
- Overalls or old clothes
- Sterile eye wash
- Skin barrier cream

Architectural details on buildings, such as 'drips' under sills, barges, string courses, etc. are important. Rainwater, directed in concentrated amounts to flow down a limewashed wall, will cause permanent streaking.

Application

Pre-wet dry surfaces using a weak limewash mix.

When the base coat has turned white the next coat can be applied. It is possible to apply at least two coats per day in average conditions. Application is reasonably fast as it is similar to painting with water. When the render can absorb no more, runs will appear on the wall. It is critical to catch all runs by brushing or they will appear through the following coats. Double back on work a few minutes after applying and brush in any runs, streaking or unevenness seen. Do this thoroughly.

For external use, add raw linseed oil to the mix. You will notice when it rains that the limewash colour stays more or less the same, while areas without raw linseed oil will darken considerably.

Raw linseed oil can cause brown staining and streaking if not mixed in thoroughly.

Traditionally in Ireland to make limewash even whiter a ball of blue was used, just as when washing white clothes. Salt was sometimes added but is not recommended as it holds moisture.

From observation, it looks as though coal dust and at times mud were used also to give grey and light tan colours.

One of the beauties of limewash, particularly with earth pigments, is that it has a variable toned finish unlike the flat sameness of modern paints.

Limewashes, if applied correctly and in favourable conditions, can have a long life expectancy.

Although limewash brushes were commonly used, so were backpack or hand-held sprayers. A backpack sprayer, such as those used on vegetable crops, with the filter removed, can be very efficient.

CHAPTER 17

READING AND RECORDING BUILDINGS

We now look at the process of examining the fabric of old buildings to see what story they have to tell us, and also what changes occurred, in what sequence and when. Later we will look at simple ways of recording on paper what we see. We will then have a record of what exists and also a means of revealing features not immediately obvious from a simple observation of the building.

A record of the current state of the building and its setting is an essential pre-requisite to making any sizeable intervention.

To read an old building requires the skills of detective, archaeologist, architectural historian, architect, craftsperson and many more. It is helped by archival research and study, comparison with similar buildings, experience, and the examination of the building as found, backed up with accurate drawings. For our purpose in this section we will only consider a part of this process, the basic examination of the building as found.

Materials

Can looking at materials in the structure of the building tell us anything?

Lime mortars: Bedding mortars for stone tell us very little visually, having stayed much the same from the time first used to the early part of the twentieth century, ie, hot lime mixes with coarse sand for rubble stone which are the most common. Lime mortars with remnants of the original fuel used in their production can be carbon dated – very useful in the analysis of historic buildings.

Lime renders: Harling or wet dash, like bedding mortars, are not very useful in helping us date buildings. Smooth renders are more useful and ruled ashlar, for instance, belongs generally to the nineteenth and early twentieth centuries, and possibly the late eighteenth century. Mouldings executed in stucco around windows and doors externally almost all belong to the nineteenth and early twentieth centuries.

Plasters: Various patches of plain, undecorated internal plaster-work can be seen on earlier historic buildings and, like bedding mortars, are impossible to date visually. Some plasterwork has been dated, for example, a fifteenth-century frescoe depicting a hunting scene still survives at Holy Cross Abbey, County Tipperary. Sixteenth-century plasterwork survives at Bunratty Castle, County Clare, and Carrick-on-Suir Castle in County Tipperary.

The eighteenth century in Ireland featured some of the finest decorative plasterwork to be seen anywhere after the arrival of Paolo and Fillipo Lafranchini, Italian migrant stuccodores from the Swiss canton of Ticino in 1739. They brought with them the International Late Baroque Style, first to Carton House in County Kildare; altogether they worked in Ireland for a period of forty years. Robert West, who worked for the Lafrachini, was the first of the Irish stuccodores to imitate them, followed by Patrick Osborne and Michael Stapleton. The most creative and artistic work – carried out over a prolonged period of time and involving the hand modelling and shaping of lime putty mixes with fine aggregates such as white marble dust into life-like imitations of birds, leaves, flowers, cherubs, etc. – was executed in-situ on ceilings. Features of this in-situ work are the free or naturalistic curves and shapes and the deep undercuts, creating near three-dimensional work .

Working from pattern books, the ceilings were first marked out as a grid and the design enlarged on to the ceiling. Support for detail in high relief was made by using timber laths, wire and nails.

Except for notable examples, plain internal plasters on stone or wood lath may be difficult to date but are generally post-seventeenth century. Decorative in-situ work on ceilings mostly belongs to the eighteenth century.

Hydraulic binders

Hydraulic limes, being for the most part naturally occurring (unless influenced by firing materials and temperatures), were always in existence so they tell us little; hydraulic/pozzolanic materials tell us more. In the late eighteenth century some of the early cements were developed, the most identifiable being Roman cement which was imported into Ireland during in the nineteenth century. This is brown/terracotta in colour and was mostly used for external renders. Portland cement was developed in 1824 and became increasingly common in the second half of the nineteenth century. It is distinguished by its greyness.

Rubble stone

Rubble stone (uncoursed) in a building reveals very little and was common to all periods. If brought to courses it is possibly late eighteenth-century, or more than likely, nineteenth-century work. Skew perps (vertical joints not at right angles to horizontal beds) are generally indicative of nineteenth-century work.

Romanesque work is recognised by the use of the semi-circular arch and the style of the voussoirs cut with zig-zags, dogtooths, heads, and other shapes.

Rubble work in the Romanesque period in Ireland has, at times, distinctive characteristics such as:

- Very large stones on the face of walls, sometimes face bedded.

- Large stones laid in courses (not brought to courses), ie, stones of approximately the same height, laid one against the other and displaying long wavy, roughly horizontal beds, are to be seen in some Irish Romanesque buildings. This method allows vertical joints to be efficiently broken on the next course without the use of too many small stones.

Cut stone

Cut stone reveals far more than rubble. Romanesque, Gothic and classical styles are all distinguished by their style of mouldings, etc. and are relatively easy to identify. More subtle identification for dating purposes can be made from tool marks.

Tool marks

Twelfth and thirteenth century: Diagonal pattern left on softer stones, such as sandstones and imported oolitic limestones (Dundry from Bristol). The diagonal pattern has been variously described as being left by a saw, axe or broad chisel.

Fourteenth and fifteenth century: Light punching as a face dressing to cut stones.

Sixteenth century: The punching becomes larger and more noticeable.

Drafted margins or square chiselling (used to take a face out of twist, see Chapter 6, Stonecutting) appear near the end of the sixteenth century.

Seventeenth century: Punching is very large and easily identifiable.

Eighteenth century: Faces of cut stone variously worked to refined surface finish with broad chisels and other tools and also punched (not as distinguishable and large as the immediately-preceding centuries). Also various rustic finishes.

Nineteenth century: Every conceivable surface finish is used, many of which appear for the first time such as rock or rustic surfaces without any punch marks and refined punch-work with a long stroke. Drill holes approximately 75mm long, visible on the face of granite from the Mourne Mountains, are used to date work in that area as post 1860s when plugs and feathers were first introduced. The holes were drilled or cut using a hand-held jumper chisel struck with a hammer.

Stone types

Looking at the geology of stones does not really help us date a building except if it is not a local stone. Then its geology will reveal from how far away it came and allow us, perhaps, to trace its importation and so date the building. The most famous importations of stone into Ireland are those between the twelfth and early-fifteenth centuries from Dundry quarry near Bristol to be seen at Christchurch Cathedral, Dublin, and St. Canice's Cathedral, Kilkenny. In the same period Caen stone from Normandy in France was also imported. Both Dundry and Caen stones are oolitic limestones, cream in colour and relatively soft.

Portland stone from Dorsetshire (another oolitic limestone) was imported in the eighteenth, nineteenth and twentieth centuries. It is to be seen on many public buildings in conjunction with native Irish stones like granite and carboniferous limestone. The nineteenth century saw the importation of various red and yellow sandstones from Britain and the re-appearance of Caen stone for repairs and replacements of original dundry stone, but also for new work such as pulpits, etc.

Brickwork

Examination of clay bricks can reveal quite an amount of information.

Clay bricks appear very late in Ireland. The first known examples are from the sixteenth century at Carrick-on-Suir Manor House (1565-1570) and at Bunratty Castle (1581). It seems that brickmaking did not really begin here until the seventeenth century. Jigginstown House at Naas, Co. Kildare, built in 1637 but never completed, is one of the earliest and the largest example of a brick building in the country. In the eighteenth and nineteenth centuries brickmaking took off in a big way, particularly in Dublin. Brickmaking was carried out by travelling brick burners who would strip the top soil and work the sub soil, if suitable (if not, materials like sand could be added) into clay bricks. This was often carried out on-site or close by. Brick burning had to be banned in the central area of Dublin city during the eighteenth century for health reasons. The bricks were burnt in a clamp kiln using turf, coal or wood, depending on location. No permanent kiln was built as the green clay bricks were used to build the clamp.

Can clay bricks help us date a building?

In general but not always, bricks with small bed heights *c*. 2in (50mm) and less are sixteenth- and seventeenth-century, and possibly imported. These are seen occasionally on buildings of that period in small quantities. At seventeenth-century Jigginstown House unusually small clay bricks, measuring approximately 7in long x 1$\frac{1}{4}$in deep x 4$\frac{1}{16}$in wide, are found at the back of the fireplaces.

Common clamp-fired bricks, which were manufactured in Ireland from the seventeenth century to World War II, tell us very little but the following may be of some assistance:

- In shape they are often twisted and curved slightly.
- Dimensionally they vary enormously, from around 8in to 10in (200mm to 250mm) in length and from 2$\frac{1}{8}$in to 3in (55mm to 76mm) in depth.
- Weight varies considerably.
- Colours are from pink to red to purple and black, and from light yellow to dark brown. In the Dublin area many have a dirty yellow face with a mauve/purple body.
- The mortar bed heights are variable because of the brick shape and probably average $\frac{3}{4}$in (20mm).
- They were sometimes 'wigged', ie, washed down with venetian red and tuck-pointed with lime putty and marble dust or fine sand to imitate more accurate work (common particularly in Dublin during the eighteenth and nineteenth centuries).

- At times they are hidden behind a better face-brick and poorly bonded to same.
- Extensively used for work on fireplaces, flues, window surrounds and under windows on rubble work.

The machine-made bricks of the second half of the nineteenth century are distinctive because they are

- Sharp in appearance.
- Dimensionally accurate.
- Generally larger, ie, 9in x 3in x $4^3/_8$in.
- Consistent in colour. Usually red or yellow.
- Heavy, particularly if pressed, in which case they may have a frog, or depression in the bed.
- Marked with manufacturer's stamp on bed sometimes.

There are other materials that can be dated besides stone, brick and lime.

Wood

The science of dendrochronology is practised at Queen's University, Belfast. By matching annual growth rings visible in wood with templates of known date, it is possible to date found timbers in old buildings.

Ireland has no medieval timber buildings still surviving and only one example or part example of a timber roof at Dunsoghley Castle, north County Dublin. Timber like oak, which was once prevalent in medieval times, became depleted by the seventeenth century for the making of charcoal to produce iron and for shipbuilding. Early nineteenth-century accounts of the country describe many places as desert-like from lack of trees.

Staircases, up to the start of the eighteenth century, were stone for defensive purposes but then were built in wood. In the Georgian period staircases were built in pine, oak and mahogany. Beautiful continuous handrails twisted around circular and elliptical curves with open wells in the finer houses, while half- and quarter-space landings were seen on simpler stairs. The earliest timber ballusters in the eighteenth century are heavy in appearance, but become more slender and refined later in the century and early nineteenth century.

Doors of the early eighteenth century have either raised or sunken panels while in the late eighteenth and early nineteenth centuries flat panels were used. In the last half of the nineteenth century heavy mouldings and four panels became the norm.

Timber panelling from floor to ceiling, common in the seventeenth and early eighteenth centuries, gives way to wainscotting on the lower half of the wall.

Pitch pine from North America is recognisable by its strong pine smell, high resin content and striking grain. Imported as beams in the nineteenth century for floors of warehouses and also large trussed roofs including church buildings.

Sash windows predominated from the eighteenth century to well into the twentieth. Horns on the bottom of the top sash first appear when glass pane sizes increase in the nineteenth century with the use of larger plate glass.

Glass – handmade, either crown or cylinder, which is slightly distorted – was made up to the end of the nineteenth century. It is distinctive in appearance and unmatched by any modern imitation.

Paint, wallpaper and limewash

Old paints can now be dated using modern technology which analyses colour pigments.

Distemper paint is used for walls and ceilings from the early eighteenth century right up to the early 1960s.

Oil paints were used from the seventeenth century on, mostly to paint wood panelling, doors, windows.

Wallpaper is common by the end of the eighteenth century.

White limewash reveals nothing in terms of dating but the addition of earth pigments helps a little. The yellow/gold/tobacco colour of the copperas wash from using green crystals (ferrous sulphate) is very nineteenth-century, while the pinks and reds of Venetian red are eighteenth- or nineteenth-century.

READING THE STRUCTURE

Simply looking at the structure of a building can provide us with useful information about changes that occurred and the sequence of those changes. From this information, combined with archival source material such as maps and documents, it is sometimes possible to deduce not only the changes and their sequence but the date also on which they took place.

Vertical joints

In some cases continuous or near continuous vertical joints indicate a change has occurred since the building was first built, usually that something has been added.

Problems at times occur in deciding which section was built first and which last.

The architectural style of each may clearly indicate which is the older, for instance if a non-defensive house of classical dimensions abuts a defensive tower house showing a vertical joint it is reason-

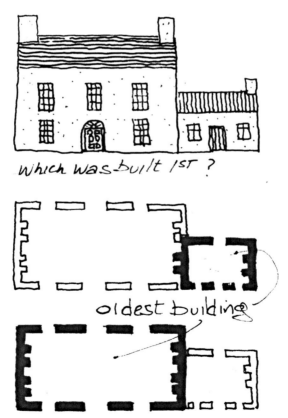

Reading buildings: party wall indicates an older section.

able to assume that the tower house was there first. At other times it is not so simple and can even elude discovery altogether.

The example on the previous page shows a two-storey farmhouse abutting a smaller single-storey house. A near continuous vertical joint is visible between the two. By looking only at the external façade it may not be possible to tell which was built first. However, by examination of the house internally we can find which building the party or dividing wall between the two belongs to, thus indicating which is the older building. (Both options shown in drawing.)

Vertical joints may show occasional horizontal stitching or bonding across the vertical joint. Where indents have been cut in the older building to 'bond in' the newer building, the mortar type of the newer building will encroach into the older building, and may at times be easily noticeable. Care needs to be taken with this; it was not an uncommon practice to leave stones projecting at ends of

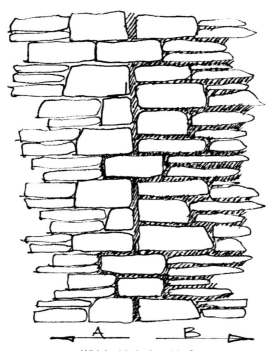

Which side is the older?
B's mortar encroaches into A's with bonding/stitching,
therefore A is the older.

buildings to allow the 'bonding in' of a future adjoined building or extension. Where this happens the mortar type of the new building will not encroach into the older building. By mortar type is meant the colour, aggregate size, texture and shape.

Vertical joints are often met with when internal plaster is missing or removed, showing original door and window openings now blocked up.

In a previous chapter on the building of a two-storey, semi-formal classical house (see Chapter 4) we saw that vertical joints commonly occurred during construction between internal and external walls and, therefore, are often contemporary with the construction of the original building and not a later addition. This, of course, varies and internal walls (particularly non-loadbearing) are often added or removed during the life of a building. In this case, ceilings may give a clue, where cornices, skirting boards or floors have been interfered with.

The widening of door and window openings, particularly the latter, is common and may be observed when the window style is inappropriate for the original style of building or there has been obvious interference with arches, cut stone or brick surrounds.

The size, shape, colour, and geology of individual stones, their tool marks if any and the quality of the masonry may all indicate later changes, but could just as easily indicate that the original source of stone ran out and the stonemason left to be replaced by somebody better or worse. Early assumptions about what is seen have often to be radically changed in the light of archival research.

RECORDING BUILDINGS

In all works of preservation, restoration or excavation, there should always be precise documentation in the form of analytical and critical reports, illustrated with drawings and photographs.
Every stage of the work of clearing, consolidation, rearrangement and integration, as well as technical and formal features identified during the course of the work, should be included. This record should be placed in the archives of a public institution and made available to research workers. It is recommended that the report should be published.

The Venice Charter 1964

This section will cover the basic recording of a building on paper. It is not aimed at buildings of regional, national or international importance but at those of humbler origin.

Before starting the measured survey check what existing documents, drawings and maps exist of the building and its surroundings. Records of this type are held in a multitude of places in Ireland (see Useful Information at the back of this book).

Reasons for recording buildings

Buildings are recorded on paper for many reasons, such as:

- To facilitate the study and understanding of the building.
- To record the building as a national monument for archival purposes.
- To record a threatened building.
- As a pre-requisite to a proposed intervention.
- In order to make an application to planning authorities, The Heritage Service or other body prior to any intervention.
- To submit along with other documents such as a specification and bill of quantities for tendering purposes.
- Part of a proposal to seek funding.
- Prior to a sale or part sale of a property or site.

Recording procedure

The measurement and recording of an existing building on paper is the reverse of the construction of a new building. This procedure involves turning a three-dimensional object into a two-dimensional one, for future reference.

Drawings in normal practice show plans, elevations and sections. Old buildings are often neither plumb, square nor level but when drawn on paper are miraculously renewed. Drawings like these are of little value because they do not interpret the building as it really exists. To avoid this happening there are a number of key points to follow:

■ Triangulate everything you possibly can both horizontally and vertically, even the pictures on the wall. This is critical – too often simply the length, width and heights of areas are measured. Triangulation at times means simply measuring the diagonals of a room or of a door or window opening. This alone results in much improved records but although we have accurately fixed the corners of these elements it does not give us the deviations in line along the face of the walls or jambs between these points. We will return to this later.

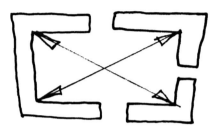

Triangulation by measurement of diagonals.

■ Use running measurements, ie, pick a zero point and measure from that point using a measuring-tape to measure off all the relevant points along the way without moving the end of the tape from zero. This limits the chances of error compared to continually moving the tape to measure every distance separately.

■ For horizontal measurements internally, choose a level if possible that is just above the level of the bottom of the windows. This means that the windows and door openings are all measured in width along this line. A metre high above floor level is normal.

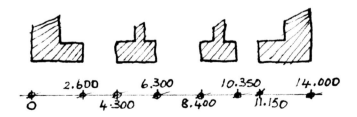

Running measurements.

On very accurate work a datum line is established. We will come back to this later.

The plan sketch with measurements – example

To begin we will take a simple one-roomed building with a single door and window. Our objective is to draw a plan of this building.

To carry out the first part of this exercise you will need an assistant and the following materials and equipment: pencil, A4 paper on a flip board, a 30 metre measuring-tape and a compass to orientate the building.

- First draw a freehand sketch of the plan of the building as best you can.
- Using running measurements from corners measure externally and internally at, say, I metre above floor level to pick up the window and, of course, the door.
- Measure both diagonals internally.
- Write down all measurements parallel to the surface measured.
- Lastly check the thickness of the walls at the door and window openings.
- Check the building in relation to the North point and mark on sketch.

We now have all we need to produce our finished drawing which is the next stage.

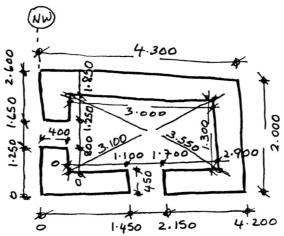

Rough sketch with measurements.

The finished plan drawing – example

The finished drawing involves using our rough sketch with measurements to produce a scaled drawing in pencil over which tracing-paper is fixed and the finished drawing produced in ink.

The following materials and equipment are needed:

A3 drawing-board incorporating a T-square and set square, A3 paper, pencil, eraser, tracing-paper (A3 @ 90 grammes per square metre), drawing pens, 0.1mm and 0.5mm scale rule, compass, our rough sketch with measurements.

The scaled pencil drawing

- We must first choose a scale. A scale of 1:100 means that the finished drawing will be one-hundredth the size of the actual building itself. Use as large a scale as possible. In this case we will use 1:50.

- Fix the A3 paper to the drawing-board and draw the scale of 1:50 on the paper in metre increments to the right of the zero point equal to the longest measurement you have to represent on the drawing. To the the left of the zero point mark out just one metre but divide this into tenths, ie, 100mm increments.

- With the pencil and the T-square draw one of the long internal sides of the building to scale.

- Now using the compass and one of the shorter internal sides as a radius measure off the scale on the drawing. Draw an arc having the point of the compass fixed at the end of one of the long sides of the building.

- At the far end of the long side and with the appropriate diagonal measurement to scale, draw another arc to cross the previous one.

- Draw a line from the long side to this point with the T-square. We now have one long and one short side of the internal shape of the building drawn in the shape of an L.

- At the far end of the long side repeat the procedure using the length of the remaining short side and diagonal. Again join the corners.

- We now only have to draw in the remaining long side. Check this for length with the scale rule, if ok then we are doing well.

- Taking the width of the walls we can now draw in the external

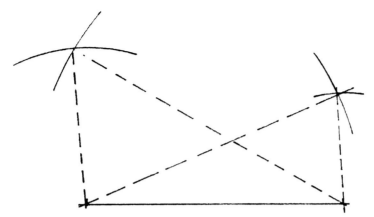

Commencement of drawing on paper with compasses.

lines of the building parallel to the inside and check these with the scale rule to see if they agree with our sketch. This is the flaw in this system – the inside is fairly accurately represented but the outside is not as accurately drawn. Sometimes a surprise awaits us: we find that a wall without any opening now reveals itself as not being the same width as the others, or the outside walls in places do not run parallel to the inside. More accurate methods are, of course, available but we are trying to keep this first exercise simple.

■ Door and window openings are now pencilled in to scale.

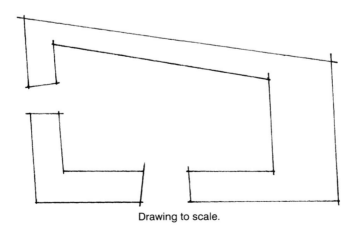

Drawing to scale.

The final ink drawing

- Now we overlay this pencilled drawing with the tracing-paper, and using pens we first draw on our scale. The scale should be kept close to the drawing of the building. This is to allow any distortion in photo-copying of the drawing to occur in the scale as well as the drawing so that they are still in agreement.

- The plan we are about to produce is, in effect, a horizontal section taken about 1 metre above floor level. This is best explained as a saw cut – as if a giant saw were used to cut the building on a horizontal line 1 metre above floor level. When we represent this cut on the finished drawing we do so with a heavy black line. Anything below this level is visible but not cut, and is represented by a lighter black line. We can also show details above this level by a dotted line.

- Working freehand (without the aid of the T-square, etc. to draw straight lines), we now draw in the lines of the building. This is done freehand to illustrate that though the building is reasonably accurate, particularly in overall dimensions and shape, the walls between corners vary slightly in line.

- In most cases, the measurements are not included on a finished drawing like this, only the scale from which somebody studying

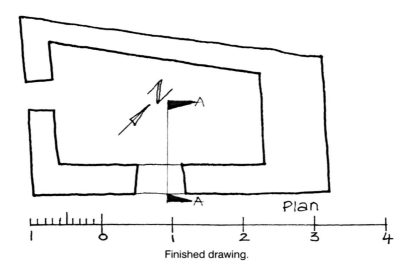

Finished drawing.

the drawing can gauge the dimensions approximately. This is done for the same reason as we produced the finished drawing freehand, to prevent the illusion of extreme accuracy.

- The North point is marked on the drawing, also the title, description, date, scale and the name of the person who completed the survey and the drawing.

Adding elevations and sections

As mentioned earlier, most drawings include not only a plan but also at least one vertical section and a number of elevations. To add a section and an elevation to our previous example we need some vertical measurements, such as the overall height of the building (ground floor to roof apex), floor to ceiling, ceiling to roof apex, window and door heights and ground to eaves height.

We can now produce both a section through the building and at least one elevation.

Section A-A of drawing on p.196.

The section is represented as a cut through the building so, as on the plan, wherever we cut is represented by a heavy black line.

We now have a drawing adequate for many types of planned intervention but not accurate enough for a true record of the building as it really exists. As explained earlier, we may have fixed our internal corners on the plan fairly accurately but not the deviations in the walls between. Equally, on both the elevation and section, wall faces, corners and jambs are all shown plumb when in fact they may lean or bulge.

For a more accurate representation of a building we must employ additional techniques as in our next example.

Accurate plans

The examples that follow use very simple, readily available equipment. More sophisticated equipment is available and used by surveyors, engineers and architects. Hand-held laser measuring devices are now affordable and very accurate. Even so, the following

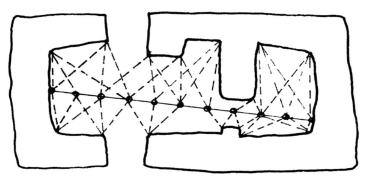

Triangulation of irregular shapes using a string line and stations.

archaic method will produce highly accurate results.

This time we will both measure and draw in one operation. To do this we require two assistants.

The finished drawing will be produced later off-site.

To carry out this exercise you will need:

    Drawing-board, paper and sticking-tape or clips
    Pencil
    Eraser
    Compass (geometric)
    Compass (directional)
    String line
    Measuring-tape (standard 30 metre)
    Scale rule
    Permanent ink pen

Our building this time is very irregular with odd angles. *Without one straight wall where does one begin?* As in so many other areas of work, like stonecutting for instance, we introduce an element that can be trusted and relied on, the straight line.

We then combine the straight line with triangles. A triangle is useful for locating points in two-dimensional space such as a drawing or a map.

Procedure

■ Stretch a string line from one end of the building to the other internally. If there are doors or windows at either end take the string line through and beyond. An ideal height is as before, I metre above floor level, but it will also work at floor level. This line should be level and secured well at both ends as it has to be pulled taut. Using the directional compass check the orientation of this line.

■ Nominate one end of the line as zero. This can be a wall or a piece of sticking-tape. From this point along the string line at regular intervals of 1, 1.5 or 2 metres fix a piece of sticking-tape to the line. Care should be taken in doing this, fixing the sticking-tape on one side to the measured interval each time. These points are stations and should be marked, 0,1,2,3, etc. with a permanent ink pen on the sticking-tape.

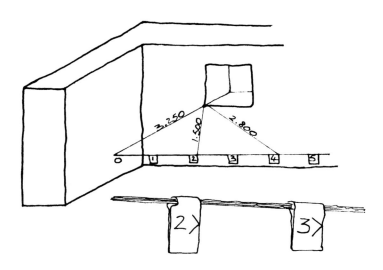

String line with sticking-tape as stations.

- Now fix the paper to the drawing-board and draw a straight line with pencil to scale with the stations ticked off along the line. Mark the scale on the drawing-paper, as in the previous example.
- From these stations you can measure triangles to any point you wish, such as window and door reveals, piers or any noticeable inconsistency along a wall face. The triangles should be similar to an equilateral shape if possible, to reduce error in measurement and in drawing.
- With three people working, one looks after the drawing, one holds the dummy end of the tape and the last person calls out the measurements. The dummy end of the tape is held on a point on the wall at the same height as the string line and stretched to a station marker, the measurement is noted and called to the draughtsperson.
- The draughtsperson working to scale now measures off this distance with the compass. Use the largest scale possible – the larger the scale the more accurate the drawing. The point of the compass is then positioned on the pencil line station that corresponds with the station chosen. An arc is now drawn with the compass.

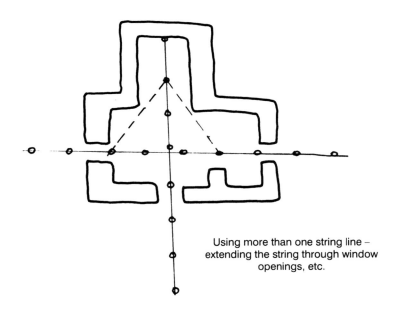

Using more than one string line – extending the string through window openings, etc.

- A second station is chosen and the measurement taken to the same point on the building. The draughtsperson does the same as before, drawing a second arc from the new station. We should now have two arcs cutting across each other on paper.
- To confirm that we are accurate (as two arcs will nearly always cut each other even when there are serious errors in measurement) a third station is chosen and a third measurement taken to the same point on the building. This is drawn on the paper as before and if we have three arcs cutting each other at the same point we are doing well. If they are seriously out then the procedure must be gone through again to find the error. This may sound a slow process, but after a short while the team of three becomes more efficient and the work can be done quite fast.
- Eventually we have a series of points either side of the straight pencil line we began with. These, when joined (if there are enough of them), will reveal on paper a close approximation of what exists in reality.
- The external face of the building can also be incorporated in this exercise, working from the same extended string line using triangulation as before. Where the string line cannot be extended through from the inside of the building to the outside, links can still usually be established with other string lines intersecting the first.
- The finished drawing is completed as before by overlaying tracing-paper on our pencilled drawing and using permanent ink pens to follow the outline free-hand, and heavy dark lines for sections, with lighter or dotted lines elsewhere.
- The scale, North point, title, description, date, and name of draughtsperson should be recorded on the finished drawing.

Datum line

What is a datum line? A datum line is an accurately established level line right around a building. It is very useful for taking accurate vertical measurements from or to other points above or below on the building which allow us to draw accurate elevations and sections. This line may be established with reference to an ordnance survey bench mark nearby (which is a fixed height above sea level, usually cut into a stone at the base of a wall or building, used for map making, building and other purposes).

In the previous example the string line we stretched from one end of the building to the other could have been fixed at a known height in relation to a datum line, or in fact be the datum line itself.

Surveying instruments like the theodolite, dumpy level or laser beam technology are mostly used to fix accurately a datum line around a building nowadays. A water level may be used on small buildings – unlike other instruments it can work around corners.

A water level is simply a long plastic tube with a glass phial at each end. Water will always seek its own level (if air bubbles don't get in the way) and so we can quickly transfer the fixed datum point around the building, marking it at set points. Later, if you wish, you can snap a chalk line between these points.

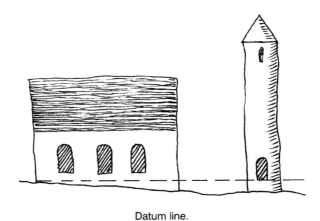

Datum line.

Accurate vertical sections and elevations

We used a horizontal string line in the previous examples from which we established various triangles in order to accurately survey the plan of a building.

For vertical sections and elevations we can do likewise by dropping plumb lines. A plumb line is a string line from which a plumb bob is attached. Plumb bobs can be suspended down stairwells, through holes in floors, out of windows and from eaves, etc. of buildings. The plumb bobs need not swing about freely but should be fixed taut at their top and bottom points once they register their plumb position. They should be as accessible as possible to the

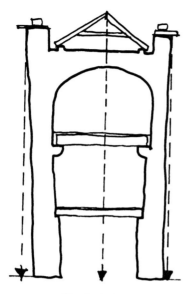

Plumbs bobs.

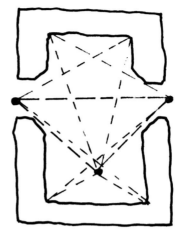

Triangulation using plumb bobs.

measuring-tape. By having at least three of these running down the height of the building it is possible to triangulate each floor level in relation to others including internal corners, wall surfaces, beams, fireplaces and other features. We can learn much from the way an old building changes structurally as it evolves upwards.

Contour gauges

Moulding details in stone, plaster or wood can be measured and drawn accurately by use of the contour gauge. The contour gauge is a series of moveable needles which will take on the shape of the object against which they are pressed. The gauge can then be placed on a sheet of paper and the detail copied full-scale. This can later be reduced to scale. It is useful when taking a mould detail to hold the gauge level, plumb or at a known angle or against some reference point so that it can later be orientated correctly on the drawing.

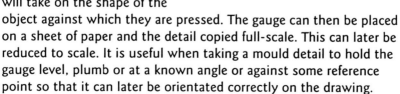

USEFUL INFORMATION

Author: Patrick McAfee
Email: mcafee@eircom.net
Website: www.irishstonewalls.com
Offers workshops, consultancy and lectures in Ireland
and elsewhere on the conservation, repair and
building of traditional stone buildings and walls.

* * *

AMSA Architectural and Monumental Stone Association
Federation House
Canal Road
Dublin 6, Ireland
Phone: (01) 497 7487

The Building Limes Forum
c/o The Schoolhouse
Rock Road
Charlestown
Fife, KY11 3EN
Scotland

Centre for Conservation
Department of Archaeology (incorporating IoAAS)
University of York
The King's Manor
York YO1 2EP
England
Phone +44 (01904) 433 901
Fax: +44 (01904) 433 902
Email: archaeology@york.ac.uk

Clogrennane Lime Ltd
Clogrennane
County Carlow
Ireland
Phone: (0503) 31811
Fax: (0503) 31607

Construction Industry Federation
Construction House
Canal Road
Dublin 6, Ireland
Phone: (01) 497 7487
Fax: (01) 496 6611

Council of Europe
Point I
F-67075 Strasbourg cedex
France
Phone international: +33 (388) 41 20 00
Fax international: +33 (388) 41 2781/82/83
Email: point i@coe.fr

Department of Conservation Sciences
Bournemouth University
Dorset House
Talbot Campus
Fen Barrow
Poole
Dorset BH12 5BB
England
Phone: +44 (0202) 595178
Fax: +44 (0202) 595255
Email: consci@bournemouth.ac.uk

Department of the Environment
Custom House
Dublin 1
Ireland
Phone: (01) 679 3377
Fax: (01) 878 6676

Department of Folklore
University College Dublin
Belfield
Dublin 4
Ireland
Phone: (01) 706 8437

Dúchas

The Heritage Service
51 St. Stephen's Green
Dublin 2, Ireland
Phone: (01) 661 3111
Dúchas combines the services of the National
Monuments, Historic Properties, National Parks
Service, Waterways Service and Wildlife Service.

Environment Service

Historic Monuments and Buildings
5-33 Hill Street
Belfast BT1 2LA
Northern Ireland
Phone: Belfast (01) 232 235000

**European Centre for the Trades and
Professions of the Conservation of
Architectural Heritage**

Isola di San Servolo
Casella Postale 676
I-30100 Venezia
Italy
Phone: +39 41 526 8546 / 526 85 47
Fax: +39 41 276 02 11
E-mail: centrove@tin.it

Feelystone

Boyle
County Roscommon
Phone: (079) 62066

Paulstown
County Kilkenny, Ireland
Phone (0503) 26191

Geological Survey of Ireland

Beggars Bush
Haddington Road
Dublin 4, Ireland
Phone (01) 670 7444

Heritage Council

Kilkenny
Co. Kilkenny, Ireland
Phone: (056) 70777
International: +353 56 70777
Fax: (056) 70788
Email: heritage@heritage.iol.ie

**IAPA Irish Association of Professional
Archaeologists**

c/o 51 St. Stephen's Green
Dublin 2, Ireland
Phone: (01) 661 3111

**ICCROM (The International Centre for
the Study of the Preservation and
Restoration of Cultural Property)**

13 Via di San Michele
I-00153 Rome, Italy
Phone international: +39 6 585 531
Fax international: +39 6 5855 3349
Email: iccrom@iccrom.org

ICOMOS

International Secretariat
49-51 Rue de la Federation
75015 Paris, France
Phone international: +33 0 145 67 67 70
Fax international: +33 0 145 66 06 22
Email: secretariat@icomos.org

Intermediate Technology

Myson House
Railway Terrace
Rugby, CV21 3HT
England
Phone: +44 (0) 1788 560631
Fax: +44 (0) 1788 540270
Email: itdg.@itdg.org.uk
Publications, overseas training and development, etc.

**IPCRA The Irish Professional
Conservators and Restorers
Association**

Phone: Belfast 381251

Irish Architectural Archive
73 Merrion Square
Dublin 2, Ireland
Phone: (01) 676 3430
Fax: (01) 661 6309
Email: iaa@archeire.com

Irish Georgian Society
Ireland's Architectural Heritage Society
74 Merrion Square
Dublin 2, Ireland
Phone: +353 1 676 7053
Fax: +353 1 662 090
Email: igs@iol.ie

McKeon Stone Ltd
Brockley Park
Stradbally
Co Laois, Ireland
Phone: (0502) 25151
Fax: (0502) 25301

Murphystone
Murphystown Road
Sandyford
Dublin 18, Ireland
Phone: (01) 295 6006
Fax: (01) 295 3694

Narrow Water Lime Service
Newry Road
Warrenpoint, Co. Down
BT34 3LE Northern Ireland
Phone: (016937) 53073
Fax: (016937) 53073

National Archives
Bishop Street
Dublin 8, Ireland
Phone: (01) 478 3711

National Building Agency
Hatherton
Richmond Avenue South
Dublin 6, Ireland
Phone: (01) 497 9654

National Library of Ireland
Kildare Street
Dublin 2, Ireland
Phone: (01) 661 8811
Fax: (01) 676 6690

**National Library of Ireland:
Department of Prints and Drawings**
Kildare Street
Dublin 2, Ireland
Phone: (01) 661 8811
Fax: (01) 676 6693

National Museum of Ireland
Kildare Street
Dublin 2, Ireland
Phone: (01) 677 7444

Public Records Office of Northern Ireland
66 Balmoral Avenue
Belfast BT 6NY
Northern Ireland
Phone: Belfast (01) 232 661 621

Registry of Deeds
Henrietta Street
Dublin 7, Ireland
Phone: (01) 873 2233

(Property transactions since 1708)

Representative Church Body Library
Braemor Park
Rathgar
Dublin 14, Ireland
Phone: (01) 492 3979
(Architectural drawings of churches etc.)

RIAI Royal Institute of Architects of Ireland
8 Merrion Square
Dublin 2, Ireland
Phone: (01) 676 1703
Fax: (01) 661 0948
Email: info@riai.ie

Roadstone Ltd
Belgard Quarry
Tallaght
Dublin 24, Ireland
Phone: (01) 452 555

Royal Irish Academy
19 Dawson Street
Dublin 2, Ireland
Phone: (01) 676 2570

Royal Society of Antiquaries of Ireland
63 Merrion Square
Dublin 2, Ireland
Phone: (01) 676 1749

School of Architecture
University College Dublin
Richview
Clonskeagh
Dublin 14, Ireland
Phone (01) 269 3244

The Scottish Lime Centre
The Schoolhouse
Rocks Road
Charlestown
Fife, KY11 3EN
Scotland

SPAB (The Society for the Protection of Ancient Buildings)
37 Spital Square
London E1 6DY
England
Phone +44 (0) 171 377 1644
Fax +44 (0) 171 247 5296
Offers membership with newsletter (SPAB News),
workshops, etc.

Stone Developments Ltd
Ballybrew
Enniskerry
County Wicklow, Ireland
Phone: (01) 286 2981
Fax: (01) 286 0449

An Taisce
The National Trust for Ireland
Tailors Hall
Back Lane
Dublin 8, Ireland
Phone: (01) 454 1786
A prescribed body under the Planning Acts, which
means it must be consulted on a variety of
planning and related matters.

Traditional Lime Company
Rath
Shillelagh Road
Tullow
County Carlow, Ireland
Phone: (0503) 51750
Fax; (0503) 52113

Trinity College
Manuscripts Department
College Green
Dublin 2, Ireland
Phone: (01) 702 1189
Fax: (01) 671 9003

Ulster Architectural Society
185 Stranmillis Road,
Belfast BT9 5DU
Northern Ireland
Phone: (01232) 660 809

UNESCO
Place de Fontenoy
75352 Paris 07SP
France
Phone international: +33 1 456 81000
Fax international: +33 1 456 71690

CHARTERS/DECLARATIONS

Charters adopted by the General Assembly of ICOMOS

- 1964 International Charter for the Conservation and Restoration of Monuments and Sites (the Venice Charter)
- 1976 Charter of Cultural Tourism
- 1982 The Florence Charter (Historic Gardens and Landscapes)
- 1987 Charter on the Conservation of Historic Towns and Urban Areas (The Washington Charter)
- 1990 Charter for the Protection and Management of the Archaeological Heritage
- 1996 Charter for the Protection and Management of the Underwater Cultural Heritage

Resolutions and Declarations of ICOMOS Symposia

- 1967 Norms of Quito (Final Report of the Meeting on the Preservation and Utilisation of Monuments and Sites of Artistic and Historical Value, Quito)
- 1972 Resolutions of the Symposium on the Introduction of Contemporary Architecture into Ancient Groups of Buildings
- 1975 Resolution on the Conservation of Smaller Historic Towns
- 1982 Tlaxcala Declaration on the Revitalisation of Small Settlements
- 1982 Declaration of Dresden
- 1983 Declaration of Rome
- 1993 Guidelines for Education and Training in the Conservation of Monuments, Ensembles and Sites
- 1994 The Nara Document on Authenticity
- 1996 Declaration of San Antonio at the Inter-American Symposium on Authenticity in the Conservation and Management of the Cultural Heritage.

Charters Adopted by ICOMOS National Committees

- 1981 The Australia ICOMOS Charter for the Conservation of Places of Cultural Significance (The Burra Charter) (Australia ICOMOS)
- 1982 Charter for the Preservation of Quebec's Heritage (Deschambault Declaration) (ICOMOS Canada)

- 1983 Appleton Charter for the Protection and Enhancement of the Built Environment (ICOMOS Canada)
- 1987 First Brazilian Seminar about the Preservation and Revitalisation of Historic Centres (ICOMOS Brazil)
- 1992 Charter for the Conservation of Places of Cultural Heritage Value (ICOMOS New Zealand)
- 1992 A Preservation Charter for the Historic Towns and Areas of the United States of America (US/ICOMOS)

Council of Europe

- European Charter of the Architectural Heritage (Council of Europe, Amsterdam, October 1975)
- Convention for the Protection of the Architectural Heritage of Europe (Council of Europe, Granada 1985)
- European Convention on the Protection of the Archaeological Heritage, revised (Council of Europe, Valetta, 1992)

Other International Standards

- Athens Charter for the Restoration of Historic Monuments (First International Congress of Architects and Technicians of Historic Monuments, Athens, 1931)
- Declaration of Amsterdam (Congress on the European Architectural Heritage, 21-25 October 1975)
- Unesco Conventions and Recommendations
- Other Cultural Protection Treaties

IRELAND

Legislation

- The National Monuments Act, 1930, amendments 1954, 1987 and 1994. Recorded monuments are protected under section 12 of the National Monuments (Amendments) Act, 1994, Archaelogical Survey of Ireland, Dúchas – The Heritage Service.
- The Local Government (Planning and Development) Acts (1963-1999) (to be replaced during 2001 by the Planning and Development Act, 2000). The Local Government (Planning and Development) Act, 1999, contains provisions for the preparation of a Record of Protected Structures within each local authority area, and contains provisions for their protection.
- The Heritage Act, 1995.

GLOSSARY

Adaptation: Modification to a building and/or its setting to allow continued use, conservation, etc. involving minimum intervention and the least possible loss of cultural heritage value.

Aggregate: Naturally occurring or machine-crushed stone, graded from coarse to fine particles. Good aggregates for rubble walls should be coarse, sharp, clean and well-graded with a 33% void or air space. In matching historic mortars the colour, size and shape of the aggregate is important.

Alumina: A compound of oxygen and aluminium and one of the constituents of clay. Associated with silica in achieving a hydraulic set in limes or cements.

Arris: The line or edge on which two surfaces, forming an exterior angle, meet each other.

Ashlar: Finely cut stone laid with very fine joints (*c*. 3mm and less).

Bank: A bench, sometimes of stone, used to support another stone for cutting.

Basalt: An igneous stone of volcanic origin.

Batter: An inward inclination of the exterior face of a wall. When associated with castles the term talus is used instead.

Bed: A horizontal layer of mortar; the top and bottom surfaces of a block of stone; a plane of stratification in a sedimentary stone.

Boasting: The dressing of a stone face with a wide chisel.

Bond: The horizontal and vertical arrangement of stones in a wall to give structural stability and to control the placement of vertical joints.

Bond stone: A stone which travels a considerable distance from one face of the wall to the other. In normal work they are at least two-thirds the width of the wall.

Boning: The sighting and paralleling of one margin of a stone with the other to take a face of a stone 'out of twist'.

Booleying: Summer pasturing of animals in mountainous areas accompanied by owners.

Caen stone: An oolitic limestone from Normandy, France.

Calcareous: Containing lime.

Calcination: Converting limestone (calcium carbonate) to quick-lime (calcium oxide) by heat as in a kiln.

Clachan: A small group of stone houses built close together, without church or shops, and often lived in by the same extended family.

Cloch: Irish term for a stone.

Conservation: The process of caring for a place so as to safeguard its cultural heritage value.

Consolidation: The introduction of one material into or on to another in order to prolong the life of the former. This should be a reversible intervention, which means it can be undone without causing damage to the original material. In the past, problems occurred with some chemical consolidants resulting in discoloration and accelerated decay. Lime-based consolidants where appropriate can be cheap, reversible and compatible. 'Consolidation as found' is a term often applied to historic ruins using minimum-intervention techniques.

Coping: The capping or covering to a wall which prevents ingress of rainwater, prevents small stones becoming dislodged from wall tops and gives an architectural style or finish to a building. They should project from both sides of a wall and have a throating underneath to shed rainwater clear of the wall. Because they are usually a cut-stone element, in some cases they are not a local stone but one imported from some other part of the country.

Corbel: A stone which is built into and projects from a wall to support a weight, ie, timber beam, stone beam or wall.

Corbelling: The art of placing one stone on another so that they oversail to enclose a roof space or similar. A common feature of Irish architecture visible at Newgrange (c. 5000 years old), in Early Christian Irish churches and later castles.

Corefill: See Hearting.

Corefilling: The replacement of lime mortars in the heart of walls, using low-pressure or gravity-feed pumping methods.

Course: A layer of stones laid to a set height.

Cramp: A device used to connect two stones together. Iron set in molten lead was commonly used for this purpose from the eighteenth century onwards. If rusted these cramps can expand and cause stone to crack and fissure.

Cyclopean: Large blocks of stone, often face bedded if sedimentary and quite thin.

Daub: To fill out a hollow spot on the face of a wall using mortar and possibly flat pieces of brick, tile or stone. Daub also refers to mud as in wattle and daub.

Dendrochronology: Matching seasonal growth rings visible in timbers with collected known patterns in order to date buildings and objects accurately.

Dimension stone: Quality stone extracted from a quarry and cut to accurate dimensions for sills, barge stones, steps and other items.

Dog: An iron dog – see Cramp.

Dóib: Irish for mud as in *dóib bhuí* or yellow mud used for building mud houses or making mud mortars.

Double stone wall: A normal wall having two faces tied across its width with through stones and bond stones, the centre being hearted with smaller stones and lime mortar (corefill).

Dowel: Generally a round bar, nowadays marine-grade stainless steel used to connect two stones together. Sometimes, in medieval work, small stones set in a cut depression in two stones were used to hold window-jamb tracery work together. A square-shaped dowel will prevent stones rotating on the axis of the dowel.

Draft: A margin of a stone at the arris worked with a chisel *c*. 25mm wide leaving a distinctive pattern of lines. Referred to as square chiselling in Ireland.

Dressing: To finish the face of a stone to a particular pattern using an appropriate tool.

Dropped square: A square used for measuring depth from a face; it has one adjustable arm.

Dundry stone: An oolitic limestone imported into Ireland from the twelfth century. The stone was quarried at Dundry near Bristol, England.

Efflorescence: The crystallisation of salts on the face of a wall. Destructive to sub-strata, plaster and paintwork if the wall is sealed and cannot breathe.

Face stone: A stone whose face is visible on the surface of the wall.

Feidín wall: A boundary wall between fields seen in east Galway and the Aran Islands off the coast of Galway and Clare. It has small stones at the base and a single wall of larger stones on top.

Finial: The top element or piece, often decorated, on a gable, spire, Celtic cross, etc.

Gneiss: Metamorphosed granite with segregated layers of quartz, feldspar and mica.

Gobán Saor: Legendary Irish craftsman.

Granite: An igneous stone comprised of quartz, feldspar, mica and amphibole (*cloch eibher* in Irish).

Harling: See Wet dashing.

Hearting: Stone and lime mortar used to fill the centre of a wall between its two external faces.

Hot lime mix: The making of mortar by mixing quicklime (CaO) and sand together with water thereby combining the slaking and mixing process resulting in high temperatures, thus the term hot lime mortars. These were normally allowed to cool and sour out over a period of time before using. These were by far the most common mortars used in Ireland in the past, particularly for all types of masonry. They were also used for external renders and base coats in internal plasterwork. Quicklime was also used with mud to create mortars.

Hydrated lime: Quicklime (CaO) slaked with just enough water to turn it to a powder C2(OH)2 which is then crushed and bagged.

Hydraulic lime: A lime which has the ability to set or part set in damp conditions. In the nineteenth century classified into weakly, moderately and eminently hydraulic lime. Hydraulic lime occurred from burning a limestone with a high mud content containing silica and alumina.

Limestone: A sedimentary stone laid down in sea water. Its main mineral is calcite derived from the bone structures of sea organisms and shells (*cloch aoile* in Irish).

Maintenance: Continuous protective care of a building, its contents and setting.

Mason's mitre: At moulded corners and elsewhere a mason does not cut a 45 degree mitre at corners as does a carpenter and joiner but cuts a straight joint and then returns the moulding on one side at 90° to match in with the other side.

Medieval: In Ireland the medieval period is best defined in three stages, early medieval 500-1169, medieval 1169-1400 and late medieval 1400-1600.

Metamorphic stone: Igneous or sedimentary stone changed from its original form through heat and/or pressure, ie, limestone metamorphosed to marble.

Minimum intervention: To reduce any intervention to an old building or object to one of necessity. To repair only what needs to be repaired. A selective rather than a global approach to interventions.

Mortice and tenon (or socket and tenon): Well known in joinery and carpentry work but also used with stone, eg, seen on c. 1000-year-old Irish Celtic crosses at the connection between shaft and base and also between the top of the cross and the finial.

Mica: A mineral with thin, flexible laminae having a shining lustre. Found in granite and many sedimentary stones.

Mica schist: A mudstone metamorphosed.

Naturally bedded: A stone laid on its flat sedimentary bed as it was originally laid down as sediment. In the quarry these beds may look distorted, twisted and even vertical from subsequent forces since they were first laid down. They should still be laid with their sedimentary bedding planes horizontal in the wall. In arches these beds should be at right angles to forces of compression.

Non-ferrous: Not iron, so there is no rust. Ferrous fixings are commonly found in eighteenth- and nineteenth-century work set in lead. If they rust, expansion results with bursting and cracking of stone. Non-ferrous fixings are preferred, usually stainless steel of marine grade quality.

Non-hydraulic lime: A lime that sets through the re-introduction of carbon dioxide. This process is called carbonation.

Oolitic limestone: Limestone having egg-shaped grains that form a structure similar to egg roe or eggs. Not common in Ireland. Portland stone from the south of England is one of the best known examples.

Out of twist: A surface taken out of winding to a true flat plane by boning and cutting.

Pig in the wall: When the string line at one end of the wall is at a different course height than it is at the other end, so that work is laid off parallel with the previous work.

Pinnings: Small stones, usually flat in shape, inserted into a mortar joint while the mortar is still soft. They reduce the overall area of mortar exposed to the weather and assist carbonation. Pinnings are also built-in during the building process to balance larger stones or to strike a level.

Pitching: The removal of stone to create a face or bed using a tool called a pitcher. A pitcher is the only cutting tool used in many types of rubble work. The pitcher is also used prior to using a punch, or punch and chisel, on more refined work. Pitchers today have tungsten tips.

Plugs and feathers: Used since Roman times to split stones. A series of holes is drilled first. This was originally done using a jumper chisel which had a bull-nose cutting edge but today compressed air and electricity is used. Two metal feathers are inserted into each hole, followed by one plug between each set of feathers. The plugs are struck in rotation and by expansion the feathers split the stone. Plugs and feathers are still used extensively in the stone industry today.

Point: A tool with a sharp point or end, traditionally forged and fire-sharpened, but mostly tungsten today. The point removes waste from the face of a stone.

Preservation: To keep as is, and to prevent any further deterioration.

Profile: Any straight edge such as a timber plank or specially made wooden or metal frame from which a string line can be pulled to set out and build a wall.

Punch: The punch does similar work to the point but it has a 6mm-wide cutting edge. In Ireland sometimes referred to as a shifting punch.

Quartz: A crystalline form of silicon dioxide. One of the main minerals found in granite stone.

Quartzite: A sandstone metamorphosed. Quartzites are usually very hard but can often be split easily on their visible bedding planes. Used for flooring, but also for building.

Quicklime: Also called lump lime, calcium oxide and CaO. Produced by burning limestone in a kiln to around 900° C which drives off carbon dioxide that accounts for about 45% of the weight of limestone. Limestone is a dangerous material capable of reaching very high temperatures if introduced to water. Either used directly with sand and water to produce a hot-lime mortar or run to putty first and then added to sand. It is the basis of all traditional lime mortars, renders and plasters. Sea shells will also produce quicklime if burnt in a kiln.

Quoin stone: A term applied to cornerstones which are used at the ends of walls, openings, etc. (*cloch choirnéil* or *cloch chúinne* in Irish). In most cases they are very accurately cut and often the same size dimensionally. They may project from the wall face or be flush with it depending on style. They sometimes do not reflect the local stone type and have obviously been imported into an area from somewhere else in the country.

Reconstruction: Based on research and without conjecture or recreation to introduce new and/or old materials into the fabric to return a building and its setting to a previous known state.

Restoration: Based on research and without conjecture, recreation or the achievement of unity of style, to return the existing fabric of a building and its setting to an earlier state by reassembly or the removal of accretions.

Retaining wall: A wall designed and built to resist the lateral pressure of earth.

Sandstone: A sedimentary stone laid down dry on land in desert-like conditions or in water. Its grains are derived from the decompostion of other rocks and cemented together with mud, carbonate, silica or iron (*cloch ghainimh* in Irish).

Single stone wall: A stone boundary wall only one stone in thickness.

Skew perp: A perpendicular joint which is intentionally cut off-plumb, common in some nineteenth-century work.

Slaking: The introduction of quicklime into water in a controlled way resulting in a thermic reaction giving temperatures beyond boiling point followed by the running of this mix through a sieve into a pit to produce lime putty. Slaking is a dangerous activity requiring protective equipment and training.

Souring out: An Irish term for the storage of mixed lime and sand in a damp condition for a period of time. Traditionally this was often six months.

Straight edge: A flat steel bar or length of timber having at least one edge straight and true.

Template: A stone projecting from a wall carrying the weight of a beam.

Templet: A sheet of zinc, plastic, thin plywood or similar cut to a particular shape, sometimes a mould, which is then applied to a stone using a scriber so that the stone can then be cut to that shape.

Through-stone: A stone which extends from one face of a wall to another thereby holding both faces together.

Wet dashing: Traditional Irish method of 'throwing on' a lime and sand mix to a wall in one or more coats. In Scotland it is called harling and in parts of England, roughcast.

Wigging: A Dublin term to describe pointing, usually of brickwork, with tuck pointing combined with venetian red washing of brickwork. Classified as bricklayers' work and belonging to the Ancient Guild of Incorporated Brick and Stonelayers. Whole families of wiggers once existed, passing down the skill from generation to generation.

BIBLIOGRAPHY

Aalen, F.H.A.,Whelan, Kevin and Stout, Matthew, *Atlas of the Irish Rural Landscape*, Cork University Press 1997.

Adam, Robert, *Classical Architecture*, Viking, Penguin, 1992.

Anon., *Working Man, Reminiscences of a Stonemason*, Murray 1908.

Architectural Conservation, an Irish Viewpoint, The Architectural Association of Ireland 1975.

Ashurst, John and Nicola, *Practical Building Conservation*, vols 1,2,3 & 4, English Heritage Technical Handbook, Gower Technical Press 1988.

Bell, Feely & Meaney, *Irish Blue Limestone, Property & Applications*, An Bord Trachtála & AMSA 1995.

Bland, William, *Arches, Piers, and Buttresses*, Crosby Lockwood & Son, 5th ed. 1890.

Boate, Gerald, *The Natural History of Ireland*, Dublin, 1755.

Bohnagen, Alfred, *Der Stukkateur und Gipser*, Zentralantiquariat der DDR, Leipzig 1987.

Bord Fáilte, *The National Monuments of Ireland*, 1964.

Boughton, R.V.,Winstone, A., & Seaward, J.C., *Brickwork*, Pitman & Sons 1945.

British Standard Code of Practice, *Cleaning and Surface Repair of Buildings*, British Standard Institution.

British Standard 5390:1976 *Code of Practice for Stone Masonry*.

Building Limes Forum, (the journal of) *Lime News*.

Conry, Michael, *Culm Crushers and Grinding Stones in the Barrow Valley and Castlecomer Plateau*, Conmore Press 1990.

Cordingley, R. A., *Normand's Parallel of the Orders of Architecture*, Tiranti, 6th ed 1959.

Costello, Peter, *Dublin Churches*, Gill and Macmillan 1989.

Council of Europe, *European Charter of the Architectural Heritage*, Amsterdam 1975.

Craig, Maurice, *Dublin 1660-1860*, Allen Figgis 1980.

Craig, Maurice, *The Architecture of Ireland from the Earliest Times to 1880*, B.T. Batsford Ltd/Easons & Sons 1983.

Danaher, Kevin, *Ireland's Traditional Houses*, Bord Fáilte.

Darby, Keith, *Church Roofing*, Church House Publishing 1988.

Department of Environment, Conservation Guidelines, 1996.

Dublin Heritage Group, *The Vernacular Buildings of East Fingal*, 1993.

Evans, E. Estyn, *Irish Folkways*, Routledge 1989.

Evans, E. Estyn, *Irish Heritage*, W. Tempest, Dundalgan Press 1977.

Flegg, Aubrey, *Course Notes, A Geological Background to Irish Building Stones*, Geological Survey of Ireland.

Forbes, T.J., *A Complete Guide to Bricklaying*, Austin Rodgers & Co. London 1926.

Hammond, Adam, *Practical Bricklaying*, Crosby Lockwood & Son, 14th ed 1924.

Harbison, Peter, *Ancient Irish Monuments*, Gill & Macmillan 1997.

Harbison, Peter, *Guide to the National Monuments of Ireland*, Gill & Macmillan 1970.

Historic Monuments and Buildings Branch, Department of the Environment, Northern Ireland, *The Care of Graveyards*, 1983.

Historic Scotland, *Preparation and Use of Lime Mortars*, Technical Advice Note, HMSO, Edinburgh 1995.

Hodge, J.C., *Brickwork for Apprentices*, Edward Arnold & Co. London 1946.

Holmes, Stafford & Michael Wingate, *Building with Lime*, Intermediate Technology Publications 1997.

Howe, J. Allen, *Stones and Quarries*, Pitman & Sons.

Howe, Malverd A., *Masonry*, John Wiley & Sons Inc. 1915.

Howley, James, *Follies and Garden Buildings of Ireland*, Yale Press.

ICOMOS, The International Council on Monuments and Sites, The Venice Charter 1964, The Cultural Tourism Charter 1976, The Florence Charter 1982, The Washington Charter 1986, The International Charter for Archaeological Heritage Management 1990.

Ireson, A.S., *Masonry Conservation and Repair*, Attic Books 1987.

Irish Georgian Society, *Traditional Building and Conservation Skills Register of Practitioners*, 1998.

Jackson, Patrick Wyse, *The Building Stones of Dublin*, Country House 1993.

James, John, *Chartres, The Masons who Built a Legend*, Routledge & Kegan Paul 1985.

Jestaz, Betrand, *Architecture of the Renaissance from Brunelleschi to Palladio*, Thames and Hudson, 1996

Johnson, David Newman, *The Irish Castle*, Irish Heritage Series, Easons & Son 1985.

Jordan, R. Furneaux, *Western Architecture*, Thames and Hudson, reprint1996.

Kearns, Kevin Corrigan, *Dublin's Vanishing Craftsmen, in Search of the Old Masters*, Appletree Press 1987.

Leask, Harold G., *Irish Castles and Castellated Houses*, Dundalgan Press 1986.

Leask, Harold G., *Irish Churches and Monasteries*, vols. 1,2 & 3, Dundalgan

Press (W. Tempest) 3rd ed 1987.

Leeds, W.H., *Architecture-orders*, Crosby Lockwood & Co. reprint 1880.

Loeber, Rolf, *Architects in Ireland, A Biographical Dictionary of 1600-1720*, John Murray Publications 1981.

McAfee, Patrick, *Irish Stone Walls*, The O'Brien Press, Dublin 1997.

McAfee, Patrick, *Stonewalling*, Conservation Guidelines, Department of the Environment 1996.

McDermott, Matthew J., and Brioscu, Aodhagan, *Dublin's Architectural Development 1800-1925*, Tulcamac 1988.

McDonnell, Joseph, *Irish Eighteenth Century Stuccowork and its European Sources*, The National Gallery of Ireland 1991.

McKay, W.B., *Building Construction*, vol. 2, Longmans 1958.

McMahon, Mary, *Medieval Church Sites of North Dublin, A Heritage Trail*, Office of Public Works 1991.

Mitchell, Charles F., *Building Construction and Drawing*, Batsford, 12th ed 1934.

Mitchell, Charles F., *Building Construction, Advanced*, Batsford, 11th ed.

Mitchell, Frank, and Ryan, Michael, *Reading the Irish Landscape*, Townhouse 1997.

Murphy, Seamus, *Stone Mad*, Routledge & Kegan Paul, 2nd ed 1986.

Nash, W.G., *Brickwork*, vols. 1,2 & 3, Hutchinson of London.

Nevill, W.E., *Geology of Ireland*, Allen Figgis 1963.

Newlands, James, *Carpenter's Assistant*, Studio Editions, London, reprint, 1990.

O'Brien, Jacqueline, and Guinness, Desmond, *Dublin, The Grand Tour*, Weidenfeld & Nicholson Ltd 1994.

Office of Public Works, Ireland, *Archaeological Survey of Ireland*, Recorded Monuments (for each county). Protected under section 12 of the National Monuments (Amendment) Act, 1994.

Office of Public Works, Ireland, *The Care and Conservation of Graveyards*, 1995.

O'Kelly, Claire, *Concise Guide to Newgrange*, C.O'Kelly, Ardnalee, Blackrock, County Cork 1984.

O'Maitiu, Seamus, and O'Reilly, Barry, Ballyknockan, *A Wicklow Stonecutter's Village*, Woodfield Press 1997.

Opderbecke/Wittenberger, *Der Steinmetz*, Callwey reprint of 1912.

O'Reilly, Sean D., *Irish Churches and Monasteries*, Collins Press, 1997.

Pepperell, Roy, *Stonemasonry Detailing*, Attic Books 1990.

Pevsner, Nikolaus, *Outline of European Architecture*, Penguin Books, reprint 1990.

Pfeiffer, Walter and Shaffrey, Maura, *Irish Cottages*, Artus Books 1990.

Purchase, William R., *Practical Masonry*, Attic Books, reprint of 1895.

Rea, J. T., *How to Estimate*, B. T. Batsford, 6th ed 1937.

Reid, Henry, *A Practical Treatise on Concrete and How to Make It, with Observations on the Uses of Cements, Limes and Mortars*, R. F. Spon 1873.

Richards, H.W., *Bricklaying and Brickcutting*.

Royal Institute of the Architects of Ireland (R.I.A.I.), *Guidelines for the Conservation of Buildings*, 1997.

Royal Institution of Chartered Surveyors and The National Federation of Building Trades Employers, Standard Method of Measurement of Building Works (SMM) 1970.

Ryan, Nicholas M., *Sparkling Granite, The Story of the Granite Working Peoples of the Three Rock Region of County Dublin*, Stone Publishing 1992.

Rzina, Franz, *Steinmetz Zeichen*, Bauverlag reprint 1989.

Semple, George, *A Treatise on Building in Water*, Dublin 1776.

Smith, David Shaw, *Ireland's Traditional Crafts*, Thames and Hudson, 2nd ed 1986.

Smith, H.P., *Constructional Archwork*, Crosby Lockwood & Son 1946.

Smith & Slater, *Classic Architecture*, Sampson Low, Marston, Searle & Rivington 1890.

Society for the Protection of Ancient Buildings (SPAB), 37 Spital Square, London, E1 6DY, Various Technical Pamphlets on Architectural Conservation.

Somers, Clarke & R. Englebach, *Ancient Egyptian Construction and Architecture*, Dover Publications.

Somerville-Large, Peter, *The Irish Country House*, Sinclair-Stevenson 1995.

Stalley, Roger, *The Cistercian Monasteries of Ireland, an Account of the History and Architecture of the White Monks in Ireland 1142-1540*, Yale University Press 1987.

Summerson, John, *Classical Language of Architecture*, Thames and Hudson, reprint, 1996.

Sweetman, David, 'Dating Irish Castles', *Archaeology Ireland*, vol. 6, number 4, issue number 22, Winter 1992.

Ulster Architectural Heritage Society in association with the Environment Service (Historic Monuments and Buildings) *Directory of Traditional Building Skills*.

Wainwright, W. Howard, & Raymond J. Whitrod, *Measurement of Building Work*, Hutchinson Educational, 1969.

Wainwright, W. Howard, and Whitrod, Raymond J., *Practical Builder's Estimating*, Hutchinson Educational 1970.

Walker, F., *Brickwork*, Crosby Lockwood & Son, 10th ed 1927.

Walker, *The Building Estimating Handbook*, Frank Walker & Co., Chicago, 7th ed 1931.

Ward, E.J. & Vollor, A., *Brickwork & Drainage*, George Allen & Unwin Ltd. 1945.

Warland, E. G., *Modern Practical Masonry*, Pitman & Sons 1953.

Whitton, J. B., *Geology and Scenery in Ireland*, Penguin Books 1974.

Wingate, Michael, *Small-Scale Lime-Burning*, Intermediate Technology Publications, 1985.

CPSIA information can be obtained at www.ICGtesting.com
Printed in the USA
BVOW042209050712

294383BV00002B/71/P